Optimal Control Methods for Linear Discrete-Time Economic Systems

Yasuo Murata

Optimal Control Methods for Linear Discrete-Time Economic Systems

Springer-Verlag
New York Heidelberg Berlin

Yasuo Murata
Nagoya City University
Faculty of Economics
Mizuhocho Mizuhoku
Nagoya, 467
Japan

AMS Classifications: 90A19, 90A20

Library of Congress Cataloging in Publication Data
Murata, Yasuo, 1931–
Optimal control methods for linear discrete-time economic systems.
Includes bibliographies and index.
1. Economic policy—Mathematical models.
2. Economic stabilization—Mathematical models.
3. Control theory. I. Title.
HD75.5.M87 1982 338.9′00724 82-19137

With 2 Figures.

Typeset by Computype, Inc., St. Paul, MN.
Printed and bound by R. R. Donnelley & Sons, Harrisonburg, VA.
Printed in the United States of America.

9 8 7 6 5 4 3 2 1

ISBN 0-387-90709-2 Springer-Verlag New York Heidelberg Berlin
ISBN 3-540-90709-2 Springer-Verlag Berlin Heidelberg New York

To Hiroko and Akiko

Preface

As our title reveals, we focus on optimal control methods and applications relevant to linear dynamic economic systems in discrete-time variables. We deal only with discrete cases simply because economic data are available in discrete forms, hence realistic economic policies should be established in discrete-time structures. Though many books have been written on optimal control in engineering, we see few on discrete-type optimal control. Moreover, since economic models take slightly different forms than do engineering ones, we need a comprehensive, self-contained treatment of linear optimal control applicable to discrete-time economic systems. The present work is intended to fill this need from the standpoint of contemporary macroeconomic stabilization.

The work is organized as follows. In Chapter 1 we demonstrate instrument instability in an economic stabilization problem and thereby establish the motivation for our departure into the optimal control world. Chapter 2 provides fundamental concepts and propositions for controlling linear deterministic discrete-time systems, together with some economic applications and numerical methods. Our optimal control rules are in the form of feedback from known state variables of the preceding period. When state variables are not observable or are accessible only with observation errors, we must obtain appropriate proxies for these variables, which are called "observers" in deterministic cases or "filters" in stochastic circumstances. In Chapters 3 and 4, respectively, Luenberger observers and Kalman filters are discussed, developed, and applied in various directions. Noticing that a separation principle lies between observer (or filter) and controller (cf. Sections 3.5 and 5.2), we are concerned in Chapter 5 with stochastic control methods in three types of uncertain environments and with the certainty equivalence principle. Existing macroeconomic applications of our control rules are examined in Section 5.5. Our main application, an economic stabilization problem, is found in Chapter 6.

Alternative routes for the journey through this book are shown in the following flow chart, where the reader will find the convenient short cut: Ch. 1 → Ch. 2 → Ch. 5 → Ch. 6. Whether the short or long course is taken, we hope that those readers with intermediate knowledges of economics, mathematics, and statistics will find few obstacles.

Ch. 1	(Instrument Instability)		Ch. 6	(Macroeconomic Stabilization)
↓			↑	
Ch. 2	(Deterministic Control)	→	Ch. 5	(Stochastic Control)
↓			↑	
Ch. 3	(Observers)	→	Ch. 4	(Filters)

The author appreciates the suggestions concerning related literature made by Associate Professor Shin-ichi Kamiyama, my colleague. I also express my thanks to Miss Yoshie Mizutori for her careful typing of my cumbersome manuscripts and Miss Mariko Akiyama for her efficient programming of computations.

Nagoya, Japan
July, 1982

YASUO MURATA

Contents

CHAPTER 1

Macroeconomic Policies and Instrument Instability

The typical macroeconomic stabilization problem presented in this chapter will serve as a motivation for our later discussion of optimal control methods. When, in a dynamic economic system, one tries to minimize the deviation in national income from its target value, one often finds an ever-increasing need to adjust policy instruments in order to offset the effects of past policies. This phenomenon was first recognized as an instrument instability by Holbrook (1972). We demonstrate this type of instability in a two-target–two-instrument setting, as an extension of Currie's article (1976) for a one-target–two-instrument setting. After presenting a Keynesian IS-LM framework for stabilization in Section 1.1, we perform optimal policies which result in the instrument instability (Section 1.2). To avoid the instability, we first propose a combination policy (Section 1.3) and then an alternative objective function involving an instrument cost (Section 1.4). Finally, in Section 1.5, we modify the optimization problem into a standard control problem so that we come naturally to the subsequent chapters.

1.1. A Keynesian Economy under the Government Budget Constraint

Currie (1976) presented optimal stabilization policies for one target (national income) in an IS-LM economy with three alternative instruments (money, government expenditure, and bonds). We shall extend his analysis to a two-target setting (national income and price level). That is, our objective is to minimize the expectation of a weighted sum of squared deviations of target variables from their equilibrium values, i.e.,

$$\min E(y_t^2 + wp_t^2), \tag{1}$$

where y_t and p_t denote the deviations in real national income Y and price level P in period t from their equilibrium values $\overline{Y}$ and $\overline{P}$, respectively; w stands for some weight attached to p_t^2.

Our basic model is composed of Currie's equations ((2), (3), and (4)) and price determination equation (5) in addition:

$$[\text{Budget}] \qquad \Delta M_t + R_t^{-1}\Delta B_t + \tau P_t Y_t - P_t G_t - (1-\tau)B_t = 0 \tag{2}$$

$$[\text{IS}] \qquad Y_t = Y(G_t, R_t, M_t, B_t, u_{yt}) \tag{3}$$

$$[\text{LM}] \qquad M_t/P_t = M(Y_t, R_t, B_t, u_{mt}) \tag{4}$$

$$[\text{Price}] \qquad P_t = P(Y_t, M_t, u_{pt}) \tag{5}$$

where

- G real government expenditure,
- M stock of high-powered money,
- B interest payments on government bonds outstanding, in money terms,
- R nominal rate of interest,
- τ tax rate,
- u_i random disturbances, ($i = y, m, p$).

Δ symbolizes a change in variable from one period to another.

Innovation is found in (5) where P is assumed to depend on Y and M in the following manner:

$$\gamma_1 \equiv \partial P/\partial Y > 0, \qquad \gamma_2 \equiv \partial P/\partial M > 0.$$

The other equations are the ones expounded in Currie's paper; we briefly explain the equations here and define new notations. (Note that, since Currie assumes fixed price, his equations do not contain price P_t as compared with equations (2) and (3).) Equation (2) shows the government budget constraint. In (3), which represents the IS curve, impact effects on Y with respect to instruments are

$$\alpha_1 \equiv \partial Y/\partial G > 0, \qquad \alpha_3 \equiv \partial Y/\partial M > 0, \qquad \alpha_4 \equiv \partial Y/\partial B > 0.$$

By the negative slope of the IS curve, we have

$$\alpha_2 \equiv -(\partial Y/\partial R)_{\text{IS}} > 0.$$

It follows from the LM curve (equation (4)) that

$$\beta_1 \equiv \partial M/\partial Y = 1/(\partial Y/\partial M)_{\text{LM}} > 0,$$

$$\beta_2 \equiv -\partial M/\partial R = -1/(\partial R/\partial M)_{\text{LM}} > 0,$$

and portfolio balance implies

$$\beta_3 \equiv \partial M/\partial B > 0.$$

Let $\bar{Y}$, $\bar{G}$, $\bar{R}$, $\bar{M}$, and $\bar{B}$ be some equilibrium values of the corresponding variables Y, G, R, M, and B, at which the government budget is balanced and goods and money markets are cleared. The associated price level $\bar{P}$ is taken to be one. Thus we have

$$\tau\bar{Y} - \bar{G} - (1-\tau)\bar{B} = 0 \tag{2'}$$

$$\bar{Y} = Y(\bar{G}, \bar{R}, \bar{M}, \bar{B}) \tag{3'}$$

$$\bar{M} = M(\bar{Y}, \bar{R}, \bar{B}) \tag{4'}$$

$$\bar{P} = P(\bar{Y}, \bar{M}) = 1. \tag{5'}$$

Consider a situation where variables deviate from the equilibrium values, and denote the deviations by the corresponding lower-case letters:

$$y_t \equiv Y_t - \bar{Y}, \qquad g_t \equiv G_t - \bar{G}, \qquad r_t \equiv R_t - \bar{R}, \qquad m_t \equiv M_t - \bar{M},$$
$$b_t \equiv B_t - \bar{B}, \qquad p_t \equiv P_t - \bar{P}.$$

Recall that Δ symbolizes a change in variable from one period to another, e.g.

$$\Delta M_t = M_t - M_{t-1} = m_t - m_{t-1} = \Delta m_t .$$

Hence (2) in deviation form is[1]

$$\Delta m_t + \Delta b_t/\bar{R} + \tau y_t + (1-\tau)\bar{B}p_t - g_t - (1-\tau)b_t = 0. \tag{6}$$

We assume that the present situation is such that Y, M, and P in equations (3), (4), and (5), respectively, can be described in linearly approximated deviation forms[2]:

$$y_t = \alpha_1 g_t - \alpha_2 r_t + \alpha_3 m_t + \alpha_4 b_t + u_{yt} \tag{7}$$

$$m_t = \bar{M}p_t + \beta_1 y_t - \beta_2 r_t + \beta_3 b_t + u_{mt} \tag{8}$$

$$p_t = \gamma_1 y_t + \gamma_2 m_t + u_{pt} . \tag{9}$$

Combining (7) with (8) to eliminate r_t yields

$$y_t = -a_0 p_t + a_1 m_t - a_2 b_t + a_3 g_t + v_{yt} \tag{10}$$

[1] The deviation forms of $R^{-1}\Delta B$ and $\tau P_t Y_t - P_t G_t$ about the equilibrium are

$$d(R^{-1}\Delta B) = d(\Delta B)/\bar{R} + \Delta\bar{B}d(R^{-1}) = (\Delta B - \Delta\bar{B})/\bar{R} = \Delta b/\bar{R},$$

and in view of (2′) and (5′)

$$d(\tau P_t Y_t - P_t G_t) = (\tau\bar{Y} - \bar{G})p_t + \tau y_t - g_t = (1-\tau)\bar{B}p_t + \tau y_t - g_t .$$

[2] The deviation form of $M_t P_t^{-1}$ about the equilibrium is

$$d(M_t P_t^{-1}) = dM_t + \bar{M}d(P_t^{-1}) = m_t - \bar{M}p_t .$$

where

$$a_0 \equiv a_4 \alpha_2 \overline{M} > 0$$
$$a_1 \equiv a_4(\alpha_2 + \alpha_3 \beta_2) > 0$$
$$a_2 \equiv a_4(\alpha_2 \beta_3 - \alpha_4 \beta_2)$$
$$a_3 \equiv a_4 \alpha_1 \beta_2 > 0$$
$$a_4 \equiv (\beta_2 + \alpha_2 \beta_1)^{-1} > 0$$
$$v_{yt} \equiv a_4(\beta_2 u_{yt} - \alpha_2 u_{mt}).$$

Substitution of (10) for y_t into (9) then yields

$$p_t = c_1 m_t - c_2 b_t + c_3 g_t + v_{pt} \tag{11}$$

where

$$c_0 \equiv (1 + a_0 \gamma_1)^{-1} > 0$$
$$c_1 \equiv c_0(a_1 \gamma_1 + \gamma_2) > 0$$
$$c_2 \equiv c_0 a_2 \gamma_1$$
$$c_3 \equiv c_0 a_3 \gamma_1 > 0$$
$$v_{pt} \equiv c_0(\gamma_1 v_{yt} + u_{pt}).$$

Finally, following Currie (1976), we assume that v_{yt} and v_{pt} obey the first-order autoregressive schemes:

$$v_{it} = h_i v_{it-1} + e_{it}, \qquad \text{for} \quad i = y, p \tag{12a}$$

where

$$|h_i| < 1, \qquad E(e_{it}) = 0, \qquad E(e_{it}^2) = \sigma_i^2 > 0, \tag{12b}$$
$$E(e_{is} e_{it}) = 0 \qquad \text{for} \quad s \neq t,$$

and

$$E(e_{yt} e_{pt}) = 0. \tag{12c}$$

Our problem is to control the system of equations (6), (10), and (11) by means of instruments g_t, m_t, and b_t so as to accomplish the minimum loss in (1). In the next section, we consider first an optimal policy with g_t and m_t chosen as policy instruments and, secondly, one with g_t and b_t chosen as control variables. Both policies prove to incur instrument instability.

1.2. Optimal Policies by Means of Two Instruments

When we control the system by government expenditure and money, the rest instrument, bonds, becomes a residual. In this case, variable b must be eliminated from the system. Substituting (10) for b_t and b_{t-1} in (6), with

(12) taken into account, yields

$$\begin{aligned}&\left(k-\bar{R}\tau a_2\right)y_t+\left(ka_0-(1-k)a_2\bar{B}\right)p_t\\&\quad=y_{t-1}+a_0p_{t-1}+\left(ka_1+\bar{R}a_2\right)m_t-\left(a_1+\bar{R}a_2\right)m_{t-1}\\&\qquad+\left(ka_3-\bar{R}a_2\right)g_t-a_3g_{t-1}-(1-kh_y)v_{yt-1}+ke_{yt},\end{aligned}\tag{13}$$

where

$$k\equiv 1-\bar{R}(1-\tau)\qquad (0<k<1).$$

Similarly, substituting (11) for b_t and b_{t-1} in (6) and taking account of (12), we have

$$\begin{aligned}&\left(k-(1-k)c_2\bar{B}\right)p_t-\bar{R}\tau c_2y_t\\&\quad=p_{t-1}+\left(kc_1+\bar{R}c_2\right)m_t-\left(c_1+\bar{R}c_2\right)m_{t-1}\\&\qquad+\left(kc_3-\bar{R}c_2\right)g_t-c_3g_{t-1}-(1-kh_p)v_{pt-1}+ke_{pt}.\end{aligned}\tag{14}$$

The system of equations (13) and (14) is solved for (y_t, p_t) as follows:

$$\begin{pmatrix}y_t\\p_t\end{pmatrix}=AJ\begin{pmatrix}y_{t-1}\\p_{t-1}\end{pmatrix}+AL\begin{pmatrix}g_t\\m_t\end{pmatrix}-AM\begin{pmatrix}g_{t-1}\\m_{t-1}\end{pmatrix}-AN\begin{pmatrix}v_{yt-1}\\v_{pt-1}\end{pmatrix}+kA\begin{pmatrix}e_{yt}\\e_{pt}\end{pmatrix}\tag{15}$$

where

$$\begin{aligned}A&\equiv\begin{bmatrix}k-\bar{R}\tau a_2 & ka_0-(1-k)a_2\bar{B}\\-\bar{R}\tau c_2 & k-(1-k)c_2\bar{B}\end{bmatrix}^{-1}\\&=\frac{1}{kD}\begin{bmatrix}k-(1-k)c_2\bar{B} & (1-k)a_2\bar{B}-ka_0\\\bar{R}\tau c_2 & k-\bar{R}\tau a_2\end{bmatrix},\end{aligned}$$

$$J\equiv\begin{bmatrix}1 & a_0\\0 & 1\end{bmatrix},\qquad L\equiv\begin{bmatrix}ka_3-\bar{R}a_2 & ka_1+\bar{R}a_2\\kc_3-\bar{R}c_2 & kc_1+\bar{R}c_2\end{bmatrix},\qquad M\equiv\begin{bmatrix}a_3 & a_1+\bar{R}a_2\\c_3 & c_1+\bar{R}c_2\end{bmatrix},$$

$$N\equiv\begin{bmatrix}1-kh_y & 0\\0 & 1-kh_p\end{bmatrix},$$

and $D\equiv k-(1-k)c_2\bar{B}-\bar{R}\tau a_2c_0$ since $a_0c_2-a_2=-a_2c_0$. Then, from the calculation of

$$\partial E\left(y_t^2+wp_t^2\right)/\partial(g_t, m_t)=0,$$

we can derive the optimal policy:

$$\begin{pmatrix} g_t \\ m_t \end{pmatrix} = L^{-1}M\begin{pmatrix} g_{t-1} \\ m_{t-1} \end{pmatrix} - L^{-1}J\begin{pmatrix} y_{t-1} \\ p_{t-1} \end{pmatrix} + L^{-1}N\begin{pmatrix} v_{yt-1} \\ v_{pt-1} \end{pmatrix}. \tag{16}$$

The expected loss in deviations from targets $E(y_t^2 + wp_t^2)$ is now computed as follows by inserting (16) into (15) and by taking (12) into account:

$$E(y_t^2 + wp_t^2) = \frac{1}{D^2}\Big[\Big(\big(k - (1-k)c_2\bar{B}\big)^2 + w\big(\bar{R}\tau c_2\big)^2\Big)\sigma_y^2$$
$$+ \Big(\big(ka_0 - (1-k)a_2\bar{B}\big)^2 + w\big(k - \bar{R}\tau a_2\big)^2\Big)\sigma_p^2\Big]. \tag{17}$$

If national income is the sole target, then w in (17) is eliminated:

$$E(y_t^2) = \frac{1}{D^2}\Big[\big(k - (1-k)c_2\bar{B}\big)^2\sigma_y^2 + \big(ka_0 - (1-k)a_2\bar{B}\big)^2\sigma_p^2\Big]. \tag{17$'$}$$

Thus, even if price deviation is deleted from our loss function (1), the evaluated optimal objective is related to the disturbance term in our price equation.

We now examine whether the optimal policy (16) follows a stationary process. This depends on the matrix $L^{-1}M$ which is found to be

$$L^{-1}M = \begin{bmatrix} q & \dfrac{(1-k^{-1})\bar{R}a_2}{ka_3 - \bar{R}a_2} \\ 0 & k^{-1} \end{bmatrix},$$

where $q \equiv a_3/(ka_3 - \bar{R}a_2)$. Thus the characteristic equation

$$0 = |\lambda I - L^{-1}M| = \lambda^2 - (q + k^{-1})\lambda + qk^{-1} \tag{18}$$

has two real roots: $\lambda_1 = k^{-1}$ and $\lambda_2 = q$. Since λ_1 is larger than unity, our policy (16) follows an expansive process, meaning an instrument instability in the sense of Holbrook (1972). The dynamic behavior of bonds follows a similar unstable process, as seen from (6′):

$$b_t = k^{-1}b_{t-1} + k^{-1}\bar{R}\big(g_t - \Delta m_t - \tau y_t - (1-\tau)\bar{B}p_t\big). \tag{6$'$}$$

An intuitive way to prevent such an instrument instability is to include in our objective function the costs associated with instrument deviations.

Next, we consider the case where government expenditure and bonds are control instruments and money is a residual instrument. Hence, as w was in the previous case, the variable m will be eliminated:

$$\begin{pmatrix} y_t \\ p_t \end{pmatrix} = \tilde{A}J\begin{pmatrix} y_{t-1} \\ p_{t-1} \end{pmatrix} + \tilde{A}\tilde{L}\begin{pmatrix} g_t \\ b_t \end{pmatrix} - \tilde{A}\tilde{M}\begin{pmatrix} g_{t-1} \\ b_{t-1} \end{pmatrix} - \tilde{A}\tilde{N}\begin{pmatrix} v_{yt-1} \\ v_{pt-1} \end{pmatrix} + \tilde{A}\begin{pmatrix} e_{yt} \\ e_{pt} \end{pmatrix} \tag{19}$$

where J is the same J as in equation (15),

$$\tilde{A} \equiv \begin{bmatrix} 1+\tau a_1 & a_0+(1-\tau)a_1\bar{B} \\ \tau c_1 & 1+(1-\tau)c_1\bar{B} \end{bmatrix}^{-1}$$

$$= \frac{1}{\tilde{D}} \begin{bmatrix} 1+(1+\tau)c_1\bar{B} & -a_0-(1-\tau)a_1\bar{B} \\ -\tau c_1 & 1+\tau a_1 \end{bmatrix},$$

$$\tilde{L} \equiv \begin{bmatrix} a_1+a_3 & -a_2-ka_1/\bar{R} \\ c_1+c_3 & -c_2-kc_1/\bar{R} \end{bmatrix}, \qquad \tilde{M} \equiv \begin{bmatrix} a_3 & -a_2-a_1/\bar{R} \\ c_3 & -c_2-c_1/\bar{R} \end{bmatrix},$$

$$\tilde{N} \equiv \begin{bmatrix} 1-h_y & 0 \\ 0 & 1-h_p \end{bmatrix},$$

and $\tilde{D} \equiv 1+\tau a_1+((1-\tau)\bar{B}-\tau a_0)c_1$. Then, from the calculation of

$$\partial E(y_t^2+wp_t^2)/\partial(g_t, b_t)=0,$$

we get the optimal policy

$$\begin{pmatrix} g_t \\ b_t \end{pmatrix} = \tilde{L}^{-1}\tilde{M}\begin{pmatrix} g_{t-1} \\ b_{t-1} \end{pmatrix} - \tilde{L}^{-1}J\begin{pmatrix} y_{t-1} \\ p_{t-1} \end{pmatrix} + \tilde{L}^{-1}\tilde{N}\begin{pmatrix} v_{yt-1} \\ v_{pt-1} \end{pmatrix} \tag{20}$$

and the corresponding expected value of the objective

$$E(y_t^2+wp_t^2) = \frac{1}{\tilde{D}^2}\Big[\Big\{\big(1+(1-\tau)c_1\bar{B}\big)^2 + w\tau^2c_1^2\Big\}\sigma_y^2 + \Big\{\big(a_0+(1-\tau)a_1\bar{B}\big)^2 + w(1+\tau a_1)^2\Big\}\sigma_p^2\Big]. \tag{21}$$

In the present case as in the previous one, we see that the optimal policy (20) does not follow a stationary process, since the associated characteristic equation

$$0 = |\lambda I - \tilde{L}^{-1}\tilde{M}| = \lambda^2-(1+q)\lambda+q, \tag{22}$$

where $q \equiv a_3/(ka_3-\bar{R}a_2)$, yields two real roots, one of which is equal to unity and the other equals q.

Concluding this section, we make one remark. In (2), the government's bond interests are supposed to be prepaid. If their payment lags by one period, the equation will change to

$$\Delta M_t + R_t^{-1}\Delta B_t + \tau P_tY_t - P_tG_t - (1-\tau)B_{t-1} = 0, \tag{2†}$$

and our optimal policies will be modified accordingly. Similarly, though, we can show that this model change does not affect the unstable properties of optimal instrument behaviors discussed earlier. (Note that this type of government budget constraint ($2^\dagger$) will be adopted in Section 1.5 and throughout Chapter 6 because one-period lagged payment of bond interests seems a reasonable and tractable assumption for discrete-time control system formation.)

1.3. A Combination Policy

In the preceding section, we considered optimal policies of two control instruments chosen out of three, g_t, m_t, and b_t. Now we combine all three basic control instruments and manipulate two independent "combined" control variables to achieve the objective (1). Our combined control variables are z_t and x_t, defined as

$$z_t = \pi_1 g_t + \pi_2 m_t + \pi_3 b_t \tag{23a}$$

$$x_t = \eta_1 g_t + \eta_2 m_t + \eta_3 b_t \tag{23b}$$

where weight vectors $\pi \equiv (\pi_1, \pi_2, \pi_3)$ and $\eta \equiv (\eta_1, \eta_2, \eta_3)$ are orthonormal, i.e., denoting transposition by superscript T,

$$\pi\pi^T = 1, \qquad \eta\eta^T = 1, \qquad \pi\eta^T = 0, \tag{24}$$

from which we obtain a relationship between basic instruments and combined control variables:

$$(g_t, m_t, b_t) = (z_t, x_t)\binom{\pi}{\eta} \tag{25}$$

which, in turn, implies the linear dependence among π, η, and the vector of basic instruments (g_t, m_t, b_t). Hence we have

$$\begin{vmatrix} \pi_1 & \pi_2 & \pi_3 \\ \eta_1 & \eta_2 & \eta_3 \\ g_t & m_t & b_t \end{vmatrix} = 0 \tag{26}$$

or, equivalently,

$$(\pi_1\eta_2 - \pi_2\eta_1)b_t + (\pi_3\eta_1 - \pi_1\eta_3)m_t + (\pi_2\eta_3 - \pi_3\eta_2)g_t = 0. \tag{26$'$}$$

Thus, given two of the three basic instruments, the rest follows from (26′). Hence our combination policy is essentially of the same idea as the previous two-control-variable policies.

The system of equations (6), (10), (11) is represented in matrix form as

$$\begin{bmatrix} \tau & (1-\tau)\bar{B} \\ 1 & a_0 \\ 0 & 1 \end{bmatrix} \begin{pmatrix} y_t \\ p_t \end{pmatrix} = \begin{bmatrix} 1 & -1 & -k/\bar{R} \\ a_3 & a_1 & -a_2 \\ c_3 & c_1 & -c_2 \end{bmatrix} \begin{bmatrix} g_t \\ m_t \\ b_t \end{bmatrix} + \begin{bmatrix} 1 & 1/\bar{R} \\ 0 & 0 \\ 0 & 0 \end{bmatrix} \begin{pmatrix} m_{t-1} \\ b_{t-1} \end{pmatrix} + \begin{bmatrix} 0 \\ v_{yt} \\ v_{pt} \end{bmatrix} \tag{27}$$

into which (25) and (12) are substituted, and the result is premultiplied by

$$\begin{pmatrix} \tau^{-1} & 0 & -\tau^{-1}(1-\tau)\bar{B} \\ 0 & 0 & 1 \end{pmatrix},$$

yielding the solution of the system in terms of the combined control variables:

$$\begin{bmatrix} y_t \\ p_t \end{bmatrix} = \begin{bmatrix} \tau^{-1}\left(f_{11}-(1-\tau)\bar{B}f_{31}\right) & \tau^{-1}\left(f_{12}-(1-\tau)\bar{B}f_{32}\right) \\ f_{31} & f_{32} \end{bmatrix} \begin{bmatrix} z_t \\ x_t \end{bmatrix}$$

$$+ \begin{bmatrix} \tau^{-1} & (\bar{R}\tau)^{-1} \\ 0 & 0 \end{bmatrix} \begin{bmatrix} m_{t-1} \\ b_{t-1} \end{bmatrix} + \begin{bmatrix} -\tau^{-1}(1-\tau)\bar{B} \\ 1 \end{bmatrix} (h_p v_{pt-1} + e_{pt}), \tag{28}$$

where

$$\begin{bmatrix} f_{11} & f_{12} \\ f_{21} & f_{22} \\ f_{31} & f_{32} \end{bmatrix} \equiv \begin{bmatrix} 1 & -1 & -k/\bar{R} \\ a_3 & a_1 & -a_2 \\ c_3 & c_1 & -c_2 \end{bmatrix} \begin{bmatrix} \pi_1 & \eta_1 \\ \pi_2 & \eta_2 \\ \pi_3 & \eta_3 \end{bmatrix}. \tag{29}$$

Denoting

$$F \equiv \begin{bmatrix} \tau^{-1}\left(f_{11}-(1-\tau)\bar{B}f_{31}\right) & \tau^{-1}\left(f_{12}-(1-\tau)\bar{B}f_{32}\right) \\ f_{31} & f_{32} \end{bmatrix}, \tag{30}$$

and $C \equiv 2(1-\tau)c_0 c_1 \gamma_1 \bar{B}$, we get

$$|F| = \begin{vmatrix} \pi_1 & \pi_2 & \pi_3 \\ \eta_1 & \eta_2 & \eta_3 \\ \dfrac{c_2 + c_1 k/\bar{R} - a_2 C}{\tau} & \dfrac{c_2 - c_3 k/\bar{R}}{\tau} & \dfrac{c_1 + c_3 - a_3 C}{\tau} \end{vmatrix}. \tag{31}$$

Since $|F| \neq 0$ in view of (26), we derive the optimal policy of the combined control variables by calculating

$$\partial E(y_t^2 + wp_t^2)/\partial(z_t, x_t) = 0.$$

That is,

$$\begin{pmatrix} z_t \\ x_t \end{pmatrix} = -F^{-1}\begin{pmatrix} \tau^{-1} & (\bar{R}\tau)^{-1} \\ 0 & 0 \end{pmatrix}\begin{pmatrix} m_{t-1} \\ b_{t-1} \end{pmatrix} - F^{-1}\begin{pmatrix} \tau^{-1}(1-\tau)\bar{B} \\ 1 \end{pmatrix} h_p v_{pt-1}. \tag{32}$$

The corresponding minimal loss of objective (1) becomes

$$E(y_t^2 + wp_t^2) = \left(\tau^{-1}(1-\tau)\bar{B}\right)^2 \sigma_p^2 + w\sigma_p^2. \tag{33}$$

The optimal policy in terms of basic control instruments is obtained by substituting (32) into (25):

$$\begin{bmatrix} g_t \\ m_t \\ b_t \end{bmatrix} = \frac{1}{\tau|F|}\begin{bmatrix} \psi_1 & \psi_1/\bar{R} \\ \psi_2 & \psi_2/\bar{R} \\ \psi_3 & \psi_3/\bar{R} \end{bmatrix}\left(\begin{matrix} m_{t-1} \\ b_{t-1} \end{matrix}\right) - (\pi^T, \eta^T)F^{-1}\begin{bmatrix} \frac{1-\tau}{\tau}\bar{B} \\ 1 \end{bmatrix} h_p v_{pt-1} \tag{34}$$

where

$$\psi_1 \equiv \eta_1 f_{31} - \pi_1 f_{32}, \qquad \psi_2 \equiv \eta_2 f_{31} - \pi_2 f_{32}, \qquad \psi_3 \equiv \eta_3 f_{31} - \pi_3 f_{32}.$$

Thus the characteristic roots associated with the optimal policy (34) are composed of two zero roots and the third one λ_3:

$$\lambda_3 = (\bar{R}\psi_2 + \psi_3)/(\bar{R}\tau|F|).$$

Hence, if and only if

$$|\bar{R}\psi_2 + \psi_3| < \bar{R}\tau|F|, \tag{35}$$

our optimal policy (34) will follow a stationary process. However, the inequality requirement (35) is not easy to check as we can see from the following relation:

$$\begin{aligned} \bar{R}\tau|F| = \bar{R}\psi_2 + \psi_3 + \bar{R}\,\{&(\pi_1\eta_2 - \pi_2\eta_1)(c_1 - a_3C) \\ &+ (\pi_3\eta_1 - \pi_1\eta_3)(c_2 + (1-\tau)c_3) \\ &- (\pi_2\eta_3 - \pi_3\eta_2)((1-\tau)c_1 + a_2C)\}. \end{aligned}$$

In any case, instrument stability is guaranteed only by inequality (35), which in turn depends on the initial selection of weight vectors π and η.

1.4. The Case Allowing Instrument Cost

So far we have studied the cases where our objective function contains the deviation losses associated with national income and price level only and not those associated with policy instruments. One main reason the optimal policies did not follow stationary processes is that objective (1) does not contain any costs associated with policy instrument. Here we propose to include in our objective the deviation loss associated with bond issuance. In order to avoid complexities of computation, we exclude p_t^2 instead from our objective function, and thus the new objective is set up as

$$\min E(y_t^2 + wb_t^2). \tag{36}$$

We want to manipulate control instruments g_t and m_t to achieve objective

(36), leaving price p_t as a residual. Thus, substituting (10) and (11), respectively, for p_t into (6), we solve the resulting equations for (y_t, b_t) as follows, with (12a) taken into account.

$$\begin{pmatrix} y_t \\ b_t \end{pmatrix} = A^*L^*\begin{pmatrix} g_t \\ m_t \end{pmatrix} + A^*J^*\begin{pmatrix} y_{t-1} \\ b_{t-1} \end{pmatrix} - A^*M^*\begin{pmatrix} g_{t-1} \\ m_{t-1} \end{pmatrix} - A^*N^*\begin{pmatrix} v_{yt-1} \\ v_{pt-1} \end{pmatrix}$$

$$- A^*(1-\tau)\overline{B}\begin{pmatrix} a_0^{-1}e_{yt} \\ e_{pt} \end{pmatrix} \tag{37}$$

where

$$D^* \equiv \left\{\left((1-\tau)\overline{B}\gamma_1 + \tau\right)a_2c_0 - k/\overline{R}\right\}(1-\tau)\overline{B}/a_0,$$

$$A^* \equiv \frac{1}{D^*}\begin{bmatrix} -(1-\tau)\overline{B}c_2 - k/\overline{R} & (1-\tau)\overline{B}a_2/a_0 - k/\overline{R} \\ -\tau & \tau - (1-\tau)\overline{B}/a_0 \end{bmatrix},$$

$$J^* \equiv \begin{pmatrix} 0 & 1/\overline{R} \\ 0 & 1/\overline{R} \end{pmatrix}, \qquad L^* \equiv \begin{bmatrix} 1-(1-\tau)\overline{B}a_3/a_0 & -1-(1-\tau)\overline{B}a_1/a_0 \\ 1-(1-\tau)\overline{B}c_3 & -1-(1-\tau)\overline{B}c_1 \end{bmatrix},$$

$$M^* \equiv \begin{pmatrix} 0 & -1 \\ 0 & -1 \end{pmatrix}, \qquad N^* \equiv \begin{bmatrix} (1-\tau)\overline{B}h_y/a_0 & 0 \\ 0 & (1-\tau)\overline{B}h_p \end{bmatrix}.$$

The corresponding optimal policy is derived from the calculation of

$$\partial E(y_t^2 + wb_t^2)/\partial(g_t, m_t) = 0$$

as

$$\begin{pmatrix} g_t \\ m_t \end{pmatrix} = L^{*-1}M^*\begin{pmatrix} g_{t-1} \\ m_{t-1} \end{pmatrix} - L^{*-1}J^*\begin{pmatrix} y_{t-1} \\ b_{t-1} \end{pmatrix} + L^{*-1}N^*\begin{pmatrix} v_{yt-1} \\ v_{pt-1} \end{pmatrix}. \tag{38}$$

Notice that the optimal policy is not affected by g_{t-1} and y_{t-1} in (38). The associated characteristic equation

$$0 = |\lambda I - L^{*-1}M^*| \tag{39}$$

yields two roots:

$$\lambda_1 = 0 \quad \text{and} \quad \lambda_2 = a_3/S,$$

since

$$L^{*-1}M^* = \frac{1}{S}\begin{pmatrix} 0 & a_0\gamma_2 - a_1 \\ 0 & a_3 \end{pmatrix}$$

where

$$S \equiv \left((1-\tau)\overline{B}\gamma_2 + 1\right)a_3 + a_1 - a_0\gamma_2.$$

For the optimal policy to follow a stationary process, therefore, an inequality requirement

$$a_1 > \left(a_0 - (1-\tau)\bar{B}a_3\right)\gamma_2 \tag{40}$$

is sufficient, since then we have $0 < \lambda_2 < 1$. Thus the optimal policy (38) is admissible with respect to the stable movement of instruments, provided (40) is valid.

We have demonstrated that the instrument instability phenomenon could be avoided by adopting either a combination policy or a loss function containing instrument costs in our one-period optimal control. For a long-span economic stabilization problem, incorporating instrument costs into our objective function is inevitable.

1.5. Formulation of a Standard Control System

In this section we try to reformulate the foregoing macroeconomic optimization problem into a dynamic stabilization problem suited to the conventional systems control methods. First, we modify our basic model so that the government's bond interest payments lag by one time period, and thus B_t in (2) and (3) is to be replaced by B_{t-1}. Accordingly, their deviation forms become, respectively,

$$\Delta m_t + \Delta b_t/\bar{R} + \tau y_t + (1-\tau)\bar{B}p_t - g_t - (1-\tau)b_{t-1} = 0, \tag{$6^\dagger$}$$

$$y_t = \alpha_1 g_t - \alpha_2 r_t + \alpha_3 m_t + \alpha_4 b_{t-1} + u_{yt}. \tag{$7^\dagger$}$$

Combining ($7^\dagger$) with (8) to eliminate r_t yields

$$y_t = -a_0 p_t + a_1 m_t - a_{20} b_t + a_{21} b_{t-1} + a_3 g_t + v_{yt} \tag{$10^\dagger$}$$

where $a_{20} \equiv a_4 \alpha_2 \beta_3 > 0$ and $a_{21} \equiv a_4 \alpha_4 \beta_2 > 0$. Note that $a_{20} - a_{21} = a_2$ by definition. Substituting ($10^\dagger$) into (9) yields

$$p_t = c_1 m_t - c_{20} b_t + c_{21} b_{t-1} + c_3 g_t + v_{yt} \tag{$11^\dagger$}$$

where $c_{20} \equiv c_0 a_{20} \gamma_1 > 0$ and $c_{21} \equiv c_0 a_{21} \gamma_1 > 0$. Note that $c_{20} - c_{21} = c_2$ by definition.

Now, as control variables, we choose government expenditure g_t and the change in money supply Δm_t in period t, instead of the total money stock m_t used in the previous analysis. The corresponding target variables are y_t, p_t, and b_t, while lagged variables g_{t-1} and b_{t-2} are treated as residuals. Therefore, our new dynamic system consists of ($6^\dagger$) and the first-order difference forms of ($10^\dagger$) and ($11^\dagger$), as shown in matrix form (41):

$$\begin{bmatrix} \tau & (1-\tau)\overline{B} & 1/\overline{R} \\ 1 & a_0 & a_{20} \\ 0 & 1 & c_{20} \end{bmatrix} \begin{bmatrix} y_t \\ p_t \\ b_t \end{bmatrix}$$

$$= \begin{bmatrix} 0 & 0 & 1-\tau+1/\overline{R} \\ 1 & a_0 & a_{20}+a_{21} \\ 0 & 1 & c_{20}+c_{21} \end{bmatrix} \begin{bmatrix} y_{t-1} \\ p_{t-1} \\ b_{t-1} \end{bmatrix} + \begin{bmatrix} 1 & -1 \\ a_3 & a_1 \\ c_3 & c_1 \end{bmatrix} \begin{pmatrix} g_t \\ \Delta m_t \end{pmatrix}$$

$$+ \begin{bmatrix} 0 & 0 \\ -a_3 & -a_{21} \\ -c_3 & -c_{21} \end{bmatrix} \begin{pmatrix} g_{t-1} \\ b_{t-2} \end{pmatrix} + \begin{bmatrix} 0 \\ v_{yt} - v_{yt-1} \\ v_{pt} - v_{pt-1} \end{bmatrix}. \tag{41}$$

Premultiplying (41) by the inverse of the matrix on the left-hand side, we can express the result in the following standard form:

$$x(t) = Ax(t-1) + Bv(t) + c(t) + D\xi(t), \tag{42}$$

where

$$x(t) \equiv \begin{bmatrix} y_t \\ p_t \\ b_t \end{bmatrix}, \qquad v(t) \equiv \begin{pmatrix} g_t \\ \Delta m_t \end{pmatrix}, \qquad \xi(t) \equiv \begin{bmatrix} 0 \\ v_{yt} - v_{yt-1} \\ v_{pt} - v_{pt-1} \end{bmatrix},$$

$$D \equiv \begin{bmatrix} \tau & (1-\tau)\overline{B} & 1/\overline{R} \\ 1 & a_0 & a_{20} \\ 0 & 1 & c_{20} \end{bmatrix}^{-1}$$

$$= \frac{1}{H} \begin{bmatrix} -c_0 a_{20} & 1/\overline{R} - c_{20}(1-\tau)\overline{B} & a_{20}(1-\tau)\overline{B} - a_0/\overline{R} \\ -c_{20} & \tau c_{20} & 1/\overline{R} - \tau a_{20} \\ 1 & -\tau & \tau a_0 - (1-\tau)\overline{B} \end{bmatrix},$$

$$A \equiv D \begin{bmatrix} 0 & 0 & 1-\tau+1/\overline{R} \\ 1 & a_0 & a_{20}+a_{21} \\ 0 & 1 & c_{20}+c_{21} \end{bmatrix}, \qquad B \equiv D \begin{bmatrix} 1 & -1 \\ a_3 & a_1 \\ c_3 & c_1 \end{bmatrix},$$

$$c(t) \equiv D \begin{bmatrix} 0 \\ -a_3 g_{t-1} - a_{21} b_{t-2} \\ -c_3 g_{t-1} - c_{21} b_{t-2} \end{bmatrix},$$

$$H \equiv -c_0 a_{20} \tau - c_{20}(1-\tau)\overline{B} + 1/\overline{R}.$$

The dynamic equation (42) is usually referred to as a *state-space form* of discrete type, and $x(t)$ is called a state vector. The corresponding deterministic state-space representation is

$$x(t) = Ax(t-1) + Bv(t) + c(t). \tag{43}$$

In the simpler case, where we neglect $c(t)$, (43) reduces to

$$x(t) = Ax(t-1) + Bv(t). \tag{43$'$}$$

As we formulated system (42) from our Keynesian equations, any linear discrete-time system with distributed lags can be transformed into a one-period lagged state-space form. (Refer, for example, to Murata (1977, pp. 386–387).) Thus we may limit our discussions to the optimal control of systems in state-space form.

As for objectives to be minimized, the previous functions (1) and (36) are concerned with one-period optimization. Multiperiod objective functions are their natural extensions, and thus we adopt the following forms of objective functions for a finite time-horizon β:

$$\min E\left\{x^T(\beta)\Gamma x(\beta) + \sum_{t=1}^{\beta} x^T(t-1)\Xi x(t-1)\right\} \tag{44}$$

or, more generally, including instrument costs,

$$\min E\left\{x^T(\beta)\Gamma x(\beta) + \sum_{t=1}^{\beta}\left(x^T(t-1)\Xi x(t-1) + v^T(t)\Phi v(t)\right)\right\}, \tag{45}$$

where Γ, Ξ, and Φ are assumed to be constant positive semidefinite symmetric matrices, with Φ positive definite. The weighing matrices will often take the simple shapes of nonnegative diagonal matrices as is the case with function (1) or (36).

For a similar formulation for stabilizing the macroeconomic instability problem, refer to Turnovsky (1974). Chow (1973) also argues that the purpose in using such a formulation for control is to solve problems associated with instrument instability in dynamic economic systems.

References

Chow, G. C., "Problems of economic policy from the viewpoint of optimal control," *American Economic Review*, vol. 63, pp. 825–837, 1973.

Currie, D. A., "Optimal stabilization policies and the government budget constraint," *Economica*, vol. 43, pp. 159–167, 1976.

Holbrook, R. S., "Optimal economic policy and the problem of instrument instability," *American Economic Review*, vol. 62, pp. 57–65, 1972.

Murata, Y., *Mathematics for Stability and Optimization of Economic Systems*. New York: Academic Press, 1977.

Turnovsky, S. J., "The stability properties of optimal economic policies," *American Economic Review*, vol. 64, pp. 136–148, 1974.

CHAPTER 2
Optimal Control of Linear Discrete-Time Systems

The aim of the basic optimal control methods for linear deterministic discrete-time systems provided here is to minimize quadratic costs associated with control variables (policy instruments) as well as target deviations. Our systems contain various sorts of exogenous variables and target values, particularly applicable to economic control problems. Section 2.1 deals with fundamental concepts and propositions concerning the controllability of discrete-time systems. In Section 2.2 optimal control laws for finite time-horizon problems will be established by following the optimality principle in dynamic programming; the laws for infinite-time problems are derived by extension of the terminal time. To implement the control laws, we give some numerical methods for solving discrete Riccati equations in Section 2.3. Then we apply the above optimal control to a dynamic Leontief system in Section 2.4. Finally, Section 2.5 is devoted to the derivation of an optimal control rule for a linear discrete-time inequality system over a finite time-horizon.

2.1. Fundamentals of Discrete-Time Control

One of the most fundamental problems of dynamic economic control is whether or not the economic system in question will be able to attain any given target by the manipulation of policy instruments. The static counterpart of this problem was initiated by Tinbergen (1952), and his rule of policy formation is summarized in Theorem 1.

Theorem 1 (Tinbergen Rule of Policy Formation). *Consider the static model represented as*

$$x = Fv, \tag{1}$$

where x is an n vector of target variables, v is an m vector of policy instruments (or control variables), and F is an $n \times m$ real constant matrix. System (1) *has a solution v for an arbitrary target x if and only if*

$$rk(F) = n, \tag{2}$$

where $rk(\cdot)$ denotes the rank of a matrix.

(For a proof, refer to Theorem 28 in Murata (1977, ch. 2).)

The Tinbergen rule (2) may be termed the *static controllability* condition from the viewpoint of modern system theory. Our main concern in the present section is to extend the Tinbergen rule to linear discrete-time dynamic systems, establishing some fundamental propositions relevant to dynamic controllability.

To study dynamic controllability, we shall consider the simple deterministic system modeled as a state-space form of discrete type:

$$x(t) = Ax(t-1) + Bv(t), \qquad t = 1, 2, \ldots, \beta \tag{3}$$

with initial condition

$$x(0) = x^0 \qquad \text{(constant)}, \tag{4}$$

where x denotes an n vector of state variables, v is an m vector of control variables, and A, B are real constant matrices of dimensions $n \times n$, $n \times m$, respectively. Note that, in any case, linear distributed-lag models of discrete-time systems can be transformed into the first-order systems of type (3). (Refer to Section 1.5.)

The general solution to (3) is obtained by an iterative substitution as follows.

$$x(t) = A^t x(0) + \sum_{\tau=0}^{t-1} A^{\tau} Bv(t-\tau), \qquad t = 1, 2, \ldots, \beta. \tag{5}$$

Given a fixed target $x(\beta) = x^1$ at some terminal time $\beta > 0$, we want to establish some necessary and sufficient conditions of attaining this target in the present system.

Definition 1. System (3) is said to be *state controllable* if x can reach a preassigned target vector x^1 at some time $\beta > 0$, starting from an arbitrary initial state (4), by manipulating control variables $v(\tau)$ $(1 \leqslant \tau \leqslant \beta)$, i.e., if

$$x(\beta) = A^{\beta} x^0 + \sum_{\tau=0}^{\beta-1} A^{\tau} Bv(\beta - \tau) = x^1. \tag{5'}$$

First we establish a straightforward dynamic version of the Tinbergen policy rule.

Theorem 2. *System* (3) *is state controllable if and only if the* $n \times (\beta m)$ *matrix*

$$P_\beta \equiv [B, AB, \ldots, A^{\beta-1}B] \tag{6}$$

has rank n.

PROOF. Equation (5′) may be rewritten as

$$P_\beta \begin{bmatrix} v(\beta) \\ v(\beta-1) \\ \vdots \\ v(1) \end{bmatrix} = x^1 - A^\beta x^0,$$

to which we apply Theorem 1. □

The rank requirement $rk(P_\beta) = n$ in Theorem 2 may be regarded as the *dynamic controllability* condition, as contrasted to the static controllability condition (2). To check the fulfillment of the rank condition, we need not calculate the rank of P_β in (6) but only that of its submatrix, in view of Theorem 3.

Theorem 3. *Define the* $n \times (jm)$ *matrix*

$$P_j \equiv [B, AB, A^2B, \ldots, A^{j-1}B]. \tag{6′}$$

If j *is the least integer such that*

$$rk(P_j) = rk(P_{j+1}),$$

then $rk(P_k) = rk(P_j)$ *for all integers* $k > j$, *and the* j *is said to be the controllability index of* (A, B).

(For a proof, see Murata (1977, p. 366).)

When β is larger than n, Theorem 2 will be altered into Theorem 2′, as follows.

Theorem 2′. *In the case of* $\beta > n$, *system* (3) *is state controllable if and only if the* $n \times (nm)$ *matrix*

$$P_n \equiv [B, AB, \ldots, A^{n-1}B] \tag{7}$$

has rank n *and* P_n *is called the state controllability matrix of system* (3).

PROOF. The "if" part of the statement is obvious, so we verify the "only if" part. Let system (3) be state controllable. Then $rk(P_\beta) = n$ by Theorem 2 and hence, by Theorem 3, an associated controllability index j exists such that

$$n + m > jm \geqslant n. \tag{$*$}$$

If $jm > n$, then $[AB, A^2B, \ldots, A^jB]$ contains jm linear combinations of all columns of matrix A, and these combinations are linearly dependent, implying that $rk(P_{j+1}) \geqslant rk(P_j)$ and that $rk(P_k) = rk(P_{j+1})$ for $k > j + 1$. Inequalities ($*$) mean that if $m = 1$ then $j = n$ and if $m \geqslant 2$ then $j < n$ for $n \geqslant 2$ and $j = 1$ for $n = 1$. Thus $rk(P_n) = rk(P_j) = n$. □

REMARKS. When $\beta < n$, $rk(P_n) = n$ is only a necessary condition for system (3) to be state controllable. Theorem 2 holds in all cases. The state controllability of system (3) is alternatively referred to as the controllability of (A, B). In the following, we use the rank requirement $rk(P_n) = n$ as an equivalent to the expression that (A, B) is controllable, since planning time horizon β will in many cases be larger than state vector dimension n and since otherwise we may extend terminal time freely.

Theorem 4 (Wonham, 1967). *Let A and B be $n \times n$ and $n \times m$ real constant matrices, respectively. If (A, B) is controllable, viz., if the $n \times (nm)$ matrix*

$$P \equiv \left[B, AB, A^2B, \ldots, A^{n-1}B \right] \tag{7'}$$

has full row rank, then $(A - BK, B)$ is controllable for each $m \times n$ real constant matrix K.

PROOF.

$$\begin{aligned}(A - BK)B &= AB - B(KB) \\ &= \{\text{second block of } P\} - \{\text{sums of columns of } B\}.\end{aligned}$$

Hence $rk[B, (A - BK)B] = rk[B, AB]$. Next, since

$$\begin{aligned}(A - BK)^2B &= A^2B - AB(KB) - B(K(A - BK)B) \\ &= \{\text{third block of } P\} - \{\text{sums of columns of } AB\} \\ &\quad - \{\text{sums of columns of } B\},\end{aligned}$$

$rk[B, (A - BK)B, (A - BK)^2B] = rk[B, AB, A^2B]$. Proceeding similarly to higher orders one by one, we finally have

$$rk\left[B, (A - BK)B, (A - BK)^2B, \ldots, (A - BK)^{n-1}B \right] = rk(P). \quad \square$$

Theorem 5 (Wonham, 1967). *Let A and B be $n \times n$ and $n \times m$ real constant matrices, respectively. The assertion that, by suitable choice of an $m \times n$ real matrix K, the set of eigenvalues of $A - BK$ can be made to correspond to any set of n distinct real scalars different from the eigenvalues of A is true if and only if (A, B) is controllable.*

(The proof is omitted since it parallels that of Theorem 2 in Chapter 3, *mutatis mutandis*.)

Another equivalence theorem of controllability is given next.

Theorem 6 (Hautus, 1969). *Let A and B be $n \times n$ and $n \times m$ real constant matrices, respectively. (A, B) is controllable if and only if the $n \times (n + m)$ matrix $[A - \lambda I, B]$ has rank n for every eigenvalue λ of A.*

PROOF. Stating that $[A - \lambda I, B]$ has full row rank is equivalent to stating that any row n vector y satisfying

$$y[A - \lambda I, B] = 0$$

must be null, in view of the linear independence of the rows of $[A - \lambda I, B]$.

"*Only if*" *part*. Suppose $y \neq 0$ exists such that $yA = \lambda y$, $yB = 0$. Then we have $yA^i = \lambda^i y$ and $yA^iB = \lambda^i yB = 0$ $(i = 0, 1, \ldots, n-1)$. Hence $yP = 0$, where $P \equiv [B, AB, A^2B, \ldots, A^{n-1}B]$ with $y \neq 0$, implying that P has linearly dependent rows, i.e., $rk(P) < n$.

"*If*" *part*. Suppose $rk(P) < n$. Then $x \neq 0$ exists such that

$$xA^iB = 0, \qquad i = 0, 1, \ldots, n-1. \tag{$*$}$$

Hence

$$x(A^{n-1} + c_{n-2}A^{n-2} + \cdots + c_1A + c_0I)B = 0 \tag{$**$}$$

for arbitrary numbers $c_i (i = 0, 1, \ldots, n-2)$. Let ψ be the minimal polynomial of x, i.e., ψ be a polynomial of the least degree such that $x\psi(A) = 0$. Such a polynomial has degree d with $1 \leqslant d \leqslant n-1$ in view of $(**)$. For some number λ and some polynomial f of degree $d-1$, we have $\psi(z) = (z - \lambda)f(z)$. Then defining $y \equiv xf(A)$, we have $0 = x\psi(A) = y(A - \lambda I)$ and from $(*)$ $0 = xf(A)B = yB$. □

Corollary. *(A, B) is controllable if and only if any row vector y satisfying $yA = \lambda y$ and $yB = 0$ is null for each eigenvalue λ of A.*

Some economic system may be described by the following state-space representation:

$$x(t) = Ax(t-1) + Bv(t), \qquad t = 1, 2, \ldots, \beta \tag{8a}$$

$$y(t) = Cx(t), \tag{8b}$$

with initial condition

$$x(0) = x^0 \qquad \text{(constant)}, \tag{4}$$

where y denotes an r vector of output variables with $r \leqslant n$, C is an $r \times n$ real constant matrix, and the other notations are those of (3). In the present system (8), we observe output vectors $y(t)$ for all t. Given an initial condition (4) and a preassigned target output $y(\beta) = y^1$ at some terminal time $\beta > 0$, we want to select an appropriate control vector $v(t)$ for all t to attain the target. The general solution to system (8) is

$$y(t) = CA^t x(0) + \sum_{\tau=0}^{t-1} CA^{\tau} Bv(t-\tau), \qquad t = 1, 2, \ldots, \beta. \tag{9}$$

Definition 2. System (8) is said to be *output controllable* if output vector y can reach the target $y(\beta) = y^1$ at some time $\beta > 0$, starting from an arbitrary initial condition $y(0) = Cx^0$, by manipulating control vector $v(\tau)$ $(1 \leqslant \tau \leqslant \beta)$, i.e., if

$$y(\beta) = CA^{\beta} x^0 + \sum_{\tau=0}^{\beta-1} CA^{\tau} Bv(\beta - \tau) = y^1. \tag{9$'$}$$

Theorem 7. *System* (8) *is output controllable if and only if the* $r \times (\beta m)$ *matrix*

$$Q_\beta \equiv [CB, CAB, \ldots, CA^{\beta-1}B] = CP_\beta \tag{10}$$

has rank r. Q_β *is called the output controllability matrix of system* (8).

In ordinary cases, the matrix C is supposed to have rank r. The supposition is important in relation to output controllability.

Theorem 8. *Let* P_j *be the* $n \times (jm)$ *matrix in* (6′), *and assume that* $rk(P_j) = n$ *for some* $j \leqslant \beta$. *Then system* (8) *is output controllable if and only if the* $r \times n$ *matrix* C *has rank* r.

PROOF.

"*If*" *part*. The Sylvester inequality is shown as

$$rk(C) + rk(P_\beta) - n \leqslant rk(CP_\beta) \leqslant \min(rk(C), rk(P_\beta)), \tag{$*$}$$

in which we consider $rk(C) = r \leqslant n = rk(P_j) = rk(P_\beta)$. Then $rk(CP_\beta) = r$ follows immediately.

"*Only if*" *part*. System (8) being output controllable implies $rk(CP_\beta) = r$ by Theorem 7. This and $rk(P_\beta) = rk(P_j) = n$ are taken into account in $(*)$, resulting in $rk(C) = r$. □

REMARK. If system (3) is state controllable and if the $r \times n$ matrix C has rank r, then system (8) will be output controllable by Theorem 2 and the Sylvester inequality $(*)$.

As a dual concept of controllability, we have "observability", which will be a prerequisite to the stability of optimal control system behavior discussed later.

Definition 3. System (8) is said to be *observable* if $x(0)$ is uniquely determined from output data $y(\tau)$ ($\tau = 0, 1, \ldots, \beta - 1$) for some positive integer β.

Theorem 9. *System* (8) *is observable if and only if the* $n \times (\beta r)$ *matrix*

$$R_\beta \equiv \left[C^T, A^T C^T, \ldots, (A^T)^{\beta-1} C^T \right] \tag{11}$$

has rank n, *where superscript* T *denotes transposition.*

PROOF. From (8b) and (9), we have

$$\begin{bmatrix} y(0) \\ y(1) - CBv(1) \\ y(2) - CBv(2) - CABv(1) \\ \vdots \\ y(\beta - 1) - \sum_{\tau=0}^{\beta-2} CA^\tau Bv(\beta - \tau - 1) \end{bmatrix} = \begin{bmatrix} C \\ CA \\ CA^2 \\ \vdots \\ CA^{\beta-1} \end{bmatrix} x(0),$$

to which we apply Theorem 1. □

We can establish the following two theorems as the duals of Theorem 3 and Theorem 2′, respectively.

Theorem 10. *Define the* $n \times (kr)$ *matrix*

$$R_k \equiv \left[C^T, A^T C^T, \ldots, (A^T)^{k-1} C^T \right]. \tag{11′}$$

If k *is the least integer such that*

$$rk(R_k) = rk(R_{k+1}),$$

then $rk(R_h) = rk(R_k)$ *for all* $h > k$, *and the* k *is said to be the observability index of* (A, C).

Theorem 9′. *In the case of* $\beta > n$, *system* (8) *is observable if and only if the* $n \times (nr)$ *matrix*

$$R \equiv \left[C^T, A^T C^T, \ldots, (A^T)^{n-1} C^T \right] \tag{12}$$

has rank n, *and* R *is called the observability matrix of system* (8).

For reasons similar to the remark following Theorem 2′, we use the rank requirement $rk(R) = n$ as an equivalent to the observability of system (8), which may alternatively be referred to as the observability of (A, C). In Chapter 3 we establish many essential propositions related to the observability concept. Here we give only a few theorems on the relationship between observability and controllability.

Theorem 6′. *(A, C) is observable if and only if any column vector x satisfying*

$$Ax = \lambda x \quad \text{and} \quad Cx = 0$$

is null for each eigenvalue λ of A.

(The proof is similar to that of Theorem 6.)

In view of Theorem 6′ and the corollary to Theorem 6, Hautus (1970) gives the following definition.

Definition 4. An eigenvalue λ of matrix A is called *(A, B)-controllable* if no nonzero row vector y exists such that $yA = \lambda y$ and $yB = 0$. An eigenvalue λ of matrix A is called *(A, C)-observable* if no nonzero column vector x exists such that $Ax = \lambda x$ and $Cx = 0$.

Remark. Clearly, (A, B) is controllable if and only if every eigenvalue of A is (A, B)-controllable, and (A, C) is observable if and only if every eigenvalue of A is (A, C)-observable.

Theorem 11 (Mårtensson, 1971). *Let A, B, C be $n \times n$, $n \times m$, $r \times n$ real constant matrices, respectively, and define the square matrix of order $2n$, called a Hamiltonian matrix:*

$$H \equiv \begin{pmatrix} A & -BB^T \\ -C^TC & -A^T \end{pmatrix}. \tag{13}$$

Then (1) if λ is an (A, B)-uncontrollable eigenvalue of A, a row vector $y \neq 0$ exists such that

$$(0, y)H^T = -\lambda(0, y),$$

where 0 denotes a null row vector; (2) if λ is an (A, C)-unobservable eigenvalue of A, a column vector $x \neq 0$ exists such that

$$H\begin{pmatrix} x \\ 0 \end{pmatrix} = \lambda \begin{pmatrix} x \\ 0 \end{pmatrix},$$

where 0 denotes a null column vector.

Proof. (1) Some $y \neq 0$ exists such that $yA = \lambda y$, $yB = 0$. For the y, we have $yA = \lambda y$, $yBB^T = 0$, which are consistent with

$$-y(BB^T, A) = (0, y)H^T = -\lambda(0, y).$$

(Proceed similarly for (2).) □

Theorem 11 means that if λ is an uncontrollable or unobservable eigenvalue, the λ is an eigenvalue of matrix H in (13). A stronger version for pure imaginary λ is given next.

Theorem 12 (Kučera, 1972). *An eigenvalue* λ *of matrix* H *in* (13) *exists such that* $\mathrm{Re}(\lambda) = 0$ *if and only if an* (A, B)*-uncontrollable and/or an* (A, C)*-unobservable eigenvalue* λ *of* A *exists such that* $\mathrm{Re}(\lambda) = 0$.

PROOF.

"If" part. Any eigenvalue of A which is (A, B)-uncontrollable and/or (A, C)-unobservable is an eigenvalue of H by virtue of Theorem 11.

"Only if" part. Set $\lambda = i\beta$, where $i^2 = -1$, and let $(x^T, y)^T$ be its associated right eigenvalue of H, conformable with the partition of H. Then

$$Ax - BB^T y^T = i\beta x, \tag{1*}$$

$$-C^T C x - A^T y^T = i\beta y^T. \tag{2*}$$

Premultiply (1*) by y and (2*) by x^T, and add the resulting equations on each side. Then we have

$$-x^T C^T C x - y BB^T y^T = 2\beta y x. \tag{3*}$$

Assuming $x^T = 0$ and $y \neq 0$ in (2*) and (3*) yields

$$yA = i\beta y \quad \text{and} \quad yB = 0,$$

and hence $\lambda = i\beta$ is an (A, B)-uncontrollable eigenvalue of A. Next, assuming $x^T \neq 0$ and $y = 0$ in (1*) and (3*) yields

$$Ax = i\beta x \quad \text{and} \quad Cx = 0,$$

and hence $\lambda = i\beta$ is an (A, C)-unobservable eigenvalue of A. □

For a further study on controllability, refer to Chapter 3 of Aoki (1976). Concluding this section, we provide the discrete version of Lyapunov stability theorem.

Theorem 13 (Discrete Version of Lyapunov Theorem (Kalman and Bertram, 1960)). *Every eigenvalue of a real square matrix* A *is less than one in modulus if and only if the matrix equation*

$$A^T B A - B = -Q, \qquad (\textit{for any real symmetric positive definite } Q) \tag{14}$$

has as its solution B *a symmetric positive definite matrix.*

PROOF.

"If" part. For any nonzero vector x

$$0 < x^T A^T B A x < x^T B x.$$

Thus a number ν exists such that $|\nu| < 1$ and $x^T A^T B A x = \nu^2 x^T B x$. Since a nonsingular matrix M exists for which $M^T M = B$, therefore, we have $MAx = \nu Mx$ or, premultiplying by M^{-1}, $[\nu I - A]x = 0$. Then for $x \neq 0$, we obtain

$$|\nu I - A| = 0.$$

Hence ν represents an eigenvalue of A.

"*Only if*" *part*. To say that all the eigenvalues of A are less than one in modulus is equivalent to saying that A^t approaches a zero matrix as time t goes to infinity. Then the linear discrete-time system

$$x(t+1) = Ax(t) \tag{15}$$

is convergent into zero as t goes to infinity.

For a symmetric positive definite matrix B, define a bounded quadratic form $V(x)$ with respect to system (15) as

$$V(x(t)) \equiv x^T(t)Bx(t) > 0 \qquad \text{for} \quad x(t) \neq 0,$$

and $V(x(t+1)) = x^T(t+1)Bx(t+1) = x^T(t)A^TBAx(t)$. It is necessary and sufficient for system (15) to be stable in the sense that the difference $V(x(t+1)) - V(x(t))$ is negative for any nonzero $x(t)$, viz., $[A^TBA - B]$ is negative definite. □

Remark. A bounded positive definite quadratic form $V(x(t))$ as defined in the proof is said to be a *Lyapunov function* in relation to discrete-time system (15), since if such a quadratic form exists, then the system is asymptotically stable.

2.2. Controllers for One-Period Lag Equation Systems

In Section 1.5, we introduced an objective loss function in quadratic cost form, to be minimized under some linear dynamic equation system. In particular, we were interested in a cost function that includes instrument costs together with target deviation losses. In this section we try to minimize the cost function of the form

$$J = x^T(\beta)\Gamma x(\beta) + \sum_{t=1}^{\beta} \left\{ x^T(t-1)\Xi x(t-1) + v^T(t)\Phi v(t) \right\} \tag{16}$$

in deterministic circumstances, with respect to $v(t)$ subject to a linear discrete-time system

$$x(t) = Ax(t-1) + Bv(t) + c(t) \tag{17}$$

for $t = 1, \ldots, \beta$, with given initial value of $x(0) = x_0$, where x is an n vector of state variables, v is an m vector of control variables, c is an n vector of exogenous variables, and A, B are constant matrices of appropriate dimensions. We assume in (16) that Γ, Ξ, and Φ are symmetric and positive semidefinite matrices with Φ positive definite and that superscript T denotes transposition.

In order to solve this finite-horizon minimization problem, we can apply Bellman's *principle of optimality* in dynamic programming, which may be stated as follows. (Refer to Murata (1977, p. 380).)

An optimal policy has the property that, at each point in time, the remaining decisions must constitute an optimal policy with regard to the state resulting from the preceding decisions. (Refer to Bellman (1957).)

Given ρ $(= 1,2, \ldots, \beta)$, we define

$$J(\beta,\rho) = x^T(\beta)\Gamma x(\beta) + \sum_{t=\rho}^{\beta} \{x^T(t-1)\Xi x(t-1) + v^T(t)\Phi v(t)\} \tag{18}$$

with $x(t)$ $(t = \beta, \beta - 1, \ldots, \rho)$ replaced by

$$x(t) = A^{t-\rho+1}x(\rho - 1) + \sum_{\tau=0}^{t-\rho} A^{\tau}\{Bv(t-\tau) + c(t-\tau)\}. \tag{19}$$

Then $J(\beta,\rho)$ is expressed only in terms of $x(\rho - 1)$, $v(t)$, and $c(t)$ $(t = \rho, \ldots, \beta)$. Denoting

$$f_\beta(x(\rho-1)) \equiv \min_{v(t)(t=\rho,\ldots,\beta)} J(\beta,\rho) \qquad \text{for} \quad \rho = 1,2,\ldots,\beta, \tag{20}$$

we have

$$f_\beta(x(0)) = \min_{v(1)}\left[H(1) + f_\beta(x(1))\right], \tag{21}$$

where

$$H(1) \equiv x^T(0)\Xi x(0) + v^T(1)\Phi v(1). \tag{22}$$

The more general form of (21) will be

$$f_\beta(x(0)) = \min_{v(t)(t=1,\ldots,\sigma)}\left[H(\sigma) + f_\beta(x(\sigma))\right] \qquad \text{for} \quad \sigma = 1,2,\ldots,\beta - 1, \tag{23}$$

where

$$H(\sigma) \equiv \sum_{t=1}^{\sigma} \{x^T(t-1)\Xi x(t-1) + v^T(t)\Phi v(t)\}. \tag{24}$$

Equation (23) is the fundamental relation representing the principle of optimality within the context of our problem.

Hence we determine the optimal control $v(t)$ $(t = 1,2, \ldots, \beta)$ such that $J(\beta,\rho)$ is minimized for all $\rho = 1,2, \ldots, \beta$. As for the calculation of $v(t)$, we proceed backward from $t = \beta$ to $t = 1$ one by one. First, differentiate $J(\beta, \beta)$ with respect to $v(\beta)$ and set the differential equal to zero. Then

$$B^T\Gamma\{Ax(\beta - 1) + Bv(\beta) + c(\beta)\} + \Phi v(\beta) = 0, \tag{25a}$$

which is solved for $v(\beta)$ as

$$v(\beta) = -K(\beta)x(\beta - 1) - \left[B^T\Gamma B + \Phi\right]^{-1}B^T\Gamma c(\beta), \tag{26a}$$

where

$$K(\beta) \equiv \left[B^T\Gamma B + \Phi\right]^{-1}B^T\Gamma A. \tag{27a}$$

Second, differentiate $J(\beta, \beta-1)$ with respect to $v(\beta-1)$, taking account of

$$x(\beta) = A^2x(\beta-2) + ABv(\beta-1) + Bv(\beta) + Ac(\beta-1) + c(\beta) \tag{19a}$$

$$x(\beta-1) = Ax(\beta-2) + Bv(\beta-1) + c(\beta-1), \tag{19b}$$

and set the differential equal to zero via (26a). This yields

$$B^TS(\beta-1)\{Ax(\beta-2) + c(\beta-1)\} + \left[B^TS(\beta-1)B + \Phi\right]v(\beta-1) + B^TL(\beta)\Gamma c(\beta) = 0, \tag{25b}$$

where

$$S(\beta-1) \equiv A^T\Gamma\left[A - BK(\beta)\right] + \Xi, \tag{28b}$$

$$L(\beta) \equiv \left[A - BK(\beta)\right]^T. \tag{29b}$$

Solve (25b) for $v(\beta-1)$:

$$v(\beta-1) = -K(\beta-1)x(\beta-2) - \left[B^TS(\beta-1)B + \Phi\right]^{-1}B^T \times \{S(\beta-1)c(\beta-1) + L(\beta)\Gamma c(\beta)\}, \tag{26b}$$

where

$$K(\beta-1) \equiv \left[B^TS(\beta-1)B + \Phi\right]^{-1}B^TS(\beta-1)A. \tag{27b}$$

Third, differentiate $J(\beta, \beta-2)$ with respect to $v(\beta-2)$, taking account of

$$x(\beta) = A^3x(\beta-3) + A^2\{Bv(\beta-2) + c(\beta-2)\} + A\{Bv(\beta-1) + c(\beta-1)\} + Bv(\beta) + c(\beta) \tag{19c}$$

$$x(\beta-1) = A^2x(\beta-3) + A\{Bv(\beta-2) + c(\beta-2)\} + Bv(\beta-1) + c(\beta-1) \tag{19d}$$

$$x(\beta-2) = Ax(\beta-3) + Bv(\beta-2) + c(\beta-2), \tag{19e}$$

and set the differential equal to zero via (26a) and (26b). This yields

$$B^TS(\beta-2)\{Ax(\beta-3) + c(\beta-2)\} + \left[B^TS(\beta-2)B + \Phi\right]v(\beta-2) + B^TL(\beta-1)\{S(\beta-1)c(\beta-1) + L(\beta)\Gamma c(\beta)\} = 0, \tag{25c}$$

where

$$S(\beta-2) \equiv A^TS(\beta-1)\left[A - BK(\beta-1)\right] + \Xi, \tag{28c}$$

$$L(\beta-1) \equiv \left[A - BK(\beta-1)\right]^T. \tag{29c}$$

Solve (25c) for $v(\beta-2)$.

$$\begin{aligned} v(\beta-2) = &-K(\beta-2)x(\beta-3)-\left[B^T S(\beta-2)B+\Phi\right]^{-1}B^T \\ &\times\{S(\beta-2)c(\beta-2)+L(\beta-1)S(\beta-1)c(\beta-1) \\ &+L(\beta-1)L(\beta)\Gamma c(\beta)\}, \end{aligned} \tag{26c}$$

where

$$K(\beta-2)\equiv\left[B^T S(\beta-2)B+\Phi\right]^{-1}B^T S(\beta-2)A. \tag{27c}$$

The general rule of optimal control for our problem is as follows.

Theorem 14. *For the finite-horizon problem of minimizing the cost criterion function J in (16) subject to system (17), the optimal control $v(t)$ is determined as*

$$v(t)=-K(t)x(t-1)-k(t), \qquad t=1,\ldots,\beta \tag{26}$$

for $t=1,2,\ldots,\beta$, where $K(t)$ is called the gain matrix defined as

$$K(t)\equiv\left[B^T S(t)B+\Phi\right]^{-1}B^T S(t)A, \tag{27}$$

$$\begin{aligned} k(t)\equiv\left[B^T S(t)B+\Phi\right]^{-1}B^T \Bigg\{ & S(t)c(t)+L(t+1)S(t+1)c(t+1) \\ & +L(t+1)L(t+2)S(t+2)c(t+2)+\cdots \\ & +\prod_{\tau=1}^{\beta-t}L(\beta-\tau+1)S(\beta)c(\beta)\Bigg\}, \end{aligned} \tag{30}$$

$$L(t)\equiv\left[A-BK(t)\right]^T, \qquad t=2,3,\ldots,\beta \tag{29}$$

and the so-called discrete Riccati equation (or Riccati difference equation) is

$$\begin{aligned} S(t-1) &= A^T S(t)\left[A-BK(t)\right]+\Xi \qquad (\textit{with } S(\beta)=\Gamma) \\ &= A^T\left[S(t)-S(t)B\left(B^T S(t)B+\Phi\right)^{-1}B^T S(t)\right]A+\Xi. \end{aligned} \tag{28}$$

The corresponding behavior of state variables is governed by

$$x(t)=\left[A-BK(t)\right]x(t-1)+c(t)-Bk(t), \tag{31}$$

and the associated minimum value of J becomes

$$\begin{aligned} \min_v J = x^T(0)S(0)x(0)+\sum_{t=1}^{\beta}\{ & k^T(t)\left[B^T S(t)B+\Phi\right]k(t) \\ & +c^T(t)S(t)c(t)+2c^T(t)S(t) \\ & \times\left[L^T(t)x(t-1)-Bk(t)\right]\} \end{aligned} \tag{32}$$

Proof. We prove (32). Eliminating Γ and Ξ in (16) and taking (28) into account, we have

$$J = x^T(0)S(0)x(0) + \sum_{t=1}^{\beta} \{x^T(t)S(t)x(t) + v^T(t)\Phi v(t) + x^T(t-1)A^TS(t)[BK(t) - A]x(t-1)\}. \tag{4*}$$

Next, we put

$$A^TS(t)B = K^T(t)[B^TS(t)B + \Phi], \tag{5*}$$

$$Ax(t-1) = x(t) - Bv(t) - c(t) \tag{6*}$$

on the right-hand side of (4*), obtaining

$$J = \sum_{t=1}^{\beta} \{v^T(t)\Phi v(t) + x^T(t-1)K^T(t)[B^TS(t)B + \Phi]K(t)x(t-1) + 2Z^T(t)S(t)x(t) - Z^T(t)S(t)Z(t)\} + x^T(0)S(0)x(0) \tag{7*}$$

where $Z(t) \equiv Bv(t) + c(t)$. Then, considering

$$Z^T(t)S(t)x(t) = Z^T(t)S(t)Ax(t-1) + Z^T(t)S(t)Z(t) \tag{8*}$$

and (5*), we obtain

$$J = \sum_{t=1}^{\beta} \{(v(t) + K(t)x(t-1))^T[B^TS(t)B + \Phi](v(t) + K(t)x(t-1)) + 2c^T(t)S(t)[Ax(t-1) + Bv(t)] + c^T(t)S(t)c(t)\} + x^T(0)S(0)x(0), \tag{9*}$$

into which we insert the optimal control $v(t)$ of (26) to yield (32). □

If $c(t)$ in system (17) is zero for each t and if the time horizon is extended to infinity, the optimal control rule will change into Theorem 15.

Theorem 15. *For the infinite-horizon case of minimizing the cost criterion function*

$$J' = \sum_{t=1}^{\infty} \{x^T(t-1)\Xi x(t-1) + v^T(t)\Phi v(t)\}, \tag{16'}$$

where Ξ and Φ are positive definite matrices, under the system

$$x(t) = Ax(t-1) + Bv(t), \qquad t = 1, 2, \ldots \tag{17†}$$

with given initial value $x(0) = x_0$, the optimal control rule is established as

($26^\dagger$) *on the assumption of state controllability of* ($17^\dagger$).

$$v(t) = -Kx(t-1), \tag{$26^\dagger$}$$

where K is the limit of $K(t)$ in (27) as $t \to \infty$ and is given by

$$K \equiv [B^TSB + \Phi]^{-1}B^TSA, \qquad (\textit{gain matrix}) \tag{$27^\dagger$}$$

$$S = A^TS[A - BK] + \Xi \qquad (\textit{Riccati equation})$$

$$= [A - BK]^TS[A - BK] + K^T\Phi K + \Xi. \tag{$28^\dagger$}$$

The corresponding behavior of state vector governed by

$$x(t) = [A - BK]x(t-1), \tag{$31^\dagger$}$$

is asymptotically stable, and

$$\min_v J' = x^T(0)Sx(0).$$

PROOF. (For the proof of the stability of ($31^\dagger$) in the Lyapunov sense of Theorem 13, see Murata (1977, p. 383).). Here we study the optimization over an infinite time interval by keeping terminal point β fixed and finite and by allowing time t to decrease indefinitely, following the method of Dorato and Levis (1971). In particular, we show that the matrix sequence $\{S(t)\}$ is bounded above and nondecreasing (i.e., $S(t-1) \geqslant S(t)$) as t goes to $-\infty$, and thus the sequence has a limit S. Since system ($17^\dagger$) is state controllable, a control input always exists that makes the performance index J' in (16′) finite by driving the state x to its preassigned target value in a finite number of steps and thereafter setting the input at zero. This reduces the infinite sum in (16′) to a finite sum. Hence $S(t)$ is bounded above if system ($17^\dagger$) is controllable.

Next we prove $S(t-1) \geqslant S(t)$. In view of (32), if optimal control $v(\tau)$ is defined for system ($17^\dagger$) over the interval $t \leqslant \tau \leqslant \beta$, we have the following minimum value of performance index $J(\beta, t+1) \equiv x^T(\beta)\Gamma x(\beta) + \sum_{\tau=t+1}^{\beta}\{x^T(\tau-1)\Xi x(\tau-1) + v^T(\tau)\Phi v(\tau)\}$:

$$\min_v J(\beta, t+1) = x^T(t)S(t)x(t) \leqslant J(\beta, t+1) \leqslant J(\beta, t) \tag{$32^\dagger$}$$

However, the extreme right-hand side of ($32^\dagger$) allows a shift from the interval $[t, \beta]$ to $[t+1, \beta+1]$. For the problem of minimizing $J(\beta+1, t+1)$ under system ($17^{\dagger\dagger}$) $x(\tau) = Ax(\tau-1) + Bv(\tau)$, the optimal control is $v(\tau) = -K(\tau)x(\tau-1)$. Putting $S(\beta) = \Gamma$ instead of $S(\beta+1) = \Gamma$, we must have

$$K(\tau) = [B^TS(\tau-1)B + \Phi]^{-1}B^TS(\tau-1)A, \quad \tau = t+1, t+2, \ldots, \beta+1 \tag{$27^{\dagger\dagger}$}$$

$$S(\tau-2) = A^TS(\tau-1)[A - BK(\tau)] + \Xi, \quad \tau = t+1, t+2, \ldots, \beta+1. \tag{$28^{\dagger\dagger}$}$$

Thus, eliminating Γ and Ξ in $J(\beta+1,t+1)$ with (28††) and $S(\beta)=\Gamma$ taken into account, we get

$$J(\beta+1,t+1) = x^T(t)S(t-1)x(t) + \sum_{\tau=t+1}^{\beta+1}\left\{x^T(\tau)S(\tau-1)x(\tau)+v^T(\tau)\Phi v(\tau) + x^T(\tau-1)A^TS(\tau-1)\left[BK(\tau)-A\right]x(\tau-1)\right\}.$$

Then, upon substitution of ($17^{\dagger\dagger}$) and $A^TS(\tau-1)B = K^T(\tau)[B^TS(\tau-1)B+\Phi]$, the terms in the braces on the right-hand side of the last equation will reduce to

$$\left(v^T(\tau)+x^T(\tau-1)K^T(\tau)\right)\left[B^TS(\tau-1)B+\Phi\right]\left(v(\tau)+K(\tau)x(\tau-1)\right).$$

Thus the minimum value of $J(\beta+1,t+1)$ corresponding to our optimal control becomes equal to $x^T(t)S(t-1)x(t)$. Connecting this with ($32^{\dagger}$) results in $S(t)\leqslant S(t-1)$. □

Now, considering the infinite-horizon counterpart under system (17) of the finite-horizon problem in Theorem 14, we may infer the following proposition from Theorems 14 and 15.

Theorem 15′. *For the infinite horizon case of minimizing the cost criterion function*

$$J' = \sum_{t=1}^{\infty}\left\{x^T(t-1)\Xi x(t-1)+v^T(t)\Phi v(t)\right\} \tag{16′}$$

(Ξ *and* Φ *are positive definite*) *under system* (17), *which is assumed to be state controllable, the optimal control is determined as*

$$v(t) = -Kx(t-1)-\bar{k}(t), \tag{26′}$$

where

$$K \equiv \left[B^TSB+\Phi\right]^{-1}B^TSA, \qquad (\textit{gain matrix}) \tag{27′}$$

$$\bar{k}(t) \equiv \left[B^TSB+\Phi\right]^{-1}B^T\left\{Sc(t)+LSc(t+1)+L^2Sc(t+2)+\ldots\right\}, \tag{30′}$$

$$L \equiv \left[A-BK\right]^T, \qquad (\textit{note that } L^{\infty}=0). \tag{29′}$$

The Riccati equation becomes the following algebraic equation:

$$\begin{aligned} S &= A^TS\left[A-BK\right]+\Xi \\ &= A^T\left[S-SB(B^TSB+\Phi)^{-1}B^TS\right]A+\Xi \\ &= A^T\left[S^{-1}+B\Phi^{-1}B^T\right]^{-1}A+\Xi \end{aligned} \tag{28′}$$

since S is positive definite and

$$[S^{-1} + B\Phi^{-1}B^T][S - SB(\Phi + B^TSB)^{-1}B^TS] = I.$$

The corresponding behavior of state variables is governed by

$$x(t) = [A - BK]x(t-1) + c(t) - B\bar{k}(t). \tag{31'}$$

Corollary. *Consider the infinite-horizon problem in Theorem* 15′ *with exogenous term* $c(t)$ *in* (17) *being constant* c^*. *Then the optimal control* $v(t)$ *in* (26′) *reduces to*

$$v(t) = -Kx(t-1) - k^*, \tag{26''}$$

where

$$k^* \equiv [B^TSB + \Phi]^{-1}B^T[I - L]^{-1}Sc^* \tag{30''}$$

with K, L, and S being those in (27′), (29′), *and* (28′), *respectively. Accordingly,* (31′) *reduces to*

$$x(t) = [A - BK]x(t-1) + c^* - Bk^*. \tag{31''}$$

Next we consider the cases where variables involved in the cost criterion function are replaced by deviations in the variables from their target values. In these cases we can obtain the following propositions in the same way as we obtained Theorem 14.

Theorem 16. *Letting* $a(t)$ *be a target value of* $x(t)$, *we try to minimize the cost function of the form*

$$J_a = (x(\beta) - a(\beta))^T\Gamma(x(\beta) - a(\beta)) + \sum_{t=1}^{\beta}\{(x(t-1) - a(t-1))^T\Xi(x(t-1) - a(t-1)) + v^T(t)\Phi v(t)\} \tag{33}$$

subject to (17), *with all the notations as defined before. The optimal control for this problem is*

$$v(t) = -K(t)x(t-1) - k_a(t), \qquad t = 1, \ldots, \beta \tag{34}$$

where

$$k_a(t) \equiv [B^TS(t)B + \Phi]^{-1}B^T \times \Big\{S(t)c_a(t) + L(t+1)S(t+1)c_a(t+1) + L(t+1)L(t+2) \times S(t+2)c_a(t+2) + \cdots + \prod_{\tau=1}^{\beta-t} L(\beta-\tau+1)S(\beta)c_a(\beta)\Big\}, \tag{35}$$

and

$$c_a(t) \equiv c(t) - a(t), \qquad t = 1, \ldots, \beta$$

with other notations being those in Theorem 14. *Accordingly, the corresponding behavior of state variables is*

$$x(t) = [A - BK(t)]x(t-1) + c(t) - Bk_a(t). \tag{36}$$

Theorem 16′. *Letting $b(t)$ be a target value of $v(t)$, we try to minimize the cost function of the form*

$$J_b = x^T(\beta)\Gamma x(\beta) + \sum_{t=1}^{\beta} \left\{ x^T(t-1)\Xi x(t-1) + (v(t) - b(t))^T \Phi (v(t) - b(t)) \right\} \tag{33′}$$

subject to (17), *with all the notations as previously defined. The optimal control for this problem is*

$$v(t) = -K(t)x(t-1) - k_b(t), \qquad t = 1, \ldots, \beta \tag{34′}$$

where

$$\begin{aligned} k_b(t) \equiv k(t) &+ \left[B^T S(t) B + \Phi \right]^{-1} \\ &\times \Bigg(B^T \Bigg\{ K^T(t+1)\Phi b(t+1) + L(t+1)K^T(t+2)\Phi b(t+2) \\ &+ L(t+1)L(t+2)K^T(t+3)\Phi b(t+3) \\ &+ \cdots + \prod_{\tau=1}^{\beta-t-1} L(\beta-\tau)K^T(\beta)\Phi b(\beta) \Bigg\} - \Phi b(t) \Bigg), \end{aligned} \tag{35′}$$

with other notations being those in Theorem 14. *Accordingly, the corresponding behavior of state variables becomes*

$$x(t) = [A - BK(t)]x(t-1) + c(t) - Bk_b(t). \tag{36′}$$

Combining Theorems 16 and 16′, we have the following.

Theorem 17. *Letting $a(t)$ and $b(t)$ be target values of $x(t)$ and $v(t)$, respectively, we try to minimize the cost function*

$$\begin{aligned} J_{ab} = (x(\beta) - a(\beta))^T \Gamma (x(\beta) - a(\beta)) &+ \sum_{t=1}^{\beta} \{ (x(t-1) - a(t-1))^T \Xi (x(t-1) - a(t-1)) \\ &+ (v(t) - b(t))^T \Phi (v(t) - b(t)) \} \end{aligned} \tag{37}$$

subject to (17), *with all the notations as defined before. The optimal control for this problem is*

$$v(t) = -K(t)x(t-1) - k_{ab}(t), \qquad t = 1, \ldots, \beta \tag{38}$$

where

$$k_{ab}(t) \equiv k_a(t) + k_b(t) - k(t) \tag{39}$$

with $k_a(t)$ *and* $k_b(t)$ *as in* (35) *and* (35′), *respectively, and with other notations being those in Theorem* 14.

If time horizon is extended to infinity, we may infer Theorem 17′ below from Theorem 17, just as we inferred Theorem 15′ from Theorem 14.

Theorem 17′. *For the infinite-horizon problem of minimizing*

$$J'_{ab} = \sum_{t=1}^{\infty} \{(x(t-1) - a(t-1))^T \Xi (x(t-1) - a(t-1)) + (v(t) - b(t))^T \Phi (v(t) - b(t))\}, \tag{37′}$$

where Ξ *and* Φ *are positive definite, subject to* (17) *which is assumed to be state controllable, the optimal control is*

$$v(t) = -Kx(t-1) - \bar{k}_{ab}(t), \tag{38′}$$

where

$$\bar{k}_{ab}(t) \equiv [B^T SB + \Phi]^{-1} \Big(B^T \{Sc_a(t) + LSc_a(t+1) + L^2 Sc_a(t+2) + \ldots\} - \Phi b(t) + B^T \{K^T \Phi b(t+1) + LK^T \Phi b(t+2) + L^2 K^T \Phi b(t+3) + \ldots\}\Big) \tag{39′}$$

with other notations being those in Theorem 15′.

For further discussions along these lines, the reader may refer to Kwakernaak and Sivan (1972, Ch. 6).

Before we apply the optimal control laws just established to economic control problems, we provide a practical way to obtain numerical solution to the discrete algebraic Riccati equation (28′).

2.3. Solving Discrete Riccati Equations

To implement the optimal control laws established in Section 2.2, we need to obtain numerical solutions to discrete Riccati equations. When terminal time β is finite, a discrete Riccati equation can be solved recursively in the direction from $t = \beta$ to $t = 1$; that is, starting with the terminal condition $S(\beta) = \Gamma$, we solve

$$S(t-1) = A^T\Big[S(t) - S(t)B\big(B^TS(t)B + \Phi\big)^{-1}B^TS(t)\Big]A + \Xi \quad (40)$$

for $t = \beta, \beta - 1, \ldots, 2, 1$. When the terminal time is infinite, we must have a solution S to the algebraic Riccati equation

$$S - A^T\Big[S - SB(B^TSB + \Phi)^{-1}B^TS\Big]A = \Xi. \quad (41)$$

S will be obtained as a unique steady-state solution to the difference equation

$$S(\tau) = A^T\Big[S(\tau-1) - S(\tau-1)B\big(B^TS(\tau-1)B + \Phi\big)^{-1}B^TS(\tau-1)\Big] \times A + \Xi \quad (40')$$

by solving it recursively for $\tau(\equiv \beta - t) = 1, 2, 3, \ldots, \beta$, under the condition $S(\tau = 0) = \Gamma = 0$, and by taking the limit of $S(\tau)$ as τ goes to infinity.

Though other methods exist for evaluating the numerical solution to the algebraic Riccati equation (41) (see for example Pappas, *et al.* (1980)), we confine ourselves here to a simple nonrecursive method (Vaughan (1970)), which is applicable to the case of nonsingularity of matrix A in the equation. The method is based on the relationship between optimal state trajectory $x(t)$ and its adjoint costate vector $p(t)$, so we first derive that relationship.

Since we are concerned with Riccati equations only, we may consider the problem of minimizing cost function J in (16) subject to the state-space form (42):

$$x(t) = Ax(t-1) + Bv(t), \qquad t = 1, 2, \ldots, \beta \quad (42)$$

with given initial value $x(0) = x_0$, where x and v are n and m vectors, respectively. (Note that all vectors are deemed column vectors in this section.) The optimal control rule for the problem is given by

$$v(t) = -K(t)x(t-1), \quad (43)$$

where gain matrix $K(t)$ is defined by (27), and the corresponding Riccati equation is shown by (28) in Theorem 14. Now, referring to Murata (1977, p. 384), we know that if the input $v(t)$ and the corresponding trajectory $x(t)$ are optimal, then an adjoint n vector sequence $\{p(t)\}$ exists such that

$$v(t) = -\Phi^{-1}B^Tp(t), \quad (44)$$

where $p(t)$ $(t = \beta - 1, \beta - 2, \ldots, 1, 0)$ is determined by

$$p(t) = \Xi x(t) + A^Tp(t+1) \quad (45)$$

and

$$\begin{aligned} Ax(t) &= x(t+1) - Bv(t+1) \\ &= x(t+1) + B\Phi^{-1}B^T p(t+1) \end{aligned} \tag{46}$$

with

$$p(\beta) = \Gamma x(\beta). \tag{47}$$

Put (46) and (45) together in matrix form as

$$\begin{pmatrix} A & 0 \\ -\Xi & I \end{pmatrix}\begin{pmatrix} x(t) \\ p(t) \end{pmatrix} = \begin{pmatrix} I & Q \\ 0 & A^T \end{pmatrix}\begin{pmatrix} x(t+1) \\ p(t+1) \end{pmatrix} \tag{48}$$

where $Q \equiv B\Phi^{-1}B^T$. Assuming that matrix A is nonsingular, from (48) we get

$$\begin{pmatrix} x(t) \\ p(t) \end{pmatrix} = \begin{pmatrix} A^{-1} & A^{-1}Q \\ \Xi A^{-1} & A^T + \Xi A^{-1}Q \end{pmatrix}\begin{pmatrix} x(t+1) \\ p(t+1) \end{pmatrix}, \qquad t = \beta - 1, \beta - 2, \dots, 1, 0 \tag{49}$$

since

$$\begin{pmatrix} A & 0 \\ -\Xi & I \end{pmatrix}^{-1} = \begin{pmatrix} A^{-1} & 0 \\ \Xi A^{-1} & I \end{pmatrix}.$$

Denoting $\tau \equiv \beta - t$, we convert (49) to

$$\begin{pmatrix} x(\tau) \\ p(\tau) \end{pmatrix} = H\begin{pmatrix} x(\tau-1) \\ p(\tau-1) \end{pmatrix}, \qquad \tau = 1, 2, \dots, \beta \tag{49'}$$

where

$$H \equiv \begin{pmatrix} A^{-1} & A^{-1}Q \\ \Xi A^{-1} & A^T + \Xi A^{-1}Q \end{pmatrix}. \tag{50}$$

The initial state $x(0) = x_0$ provides n boundary conditions; the remaining n boundary conditions are

$$p(\tau = 0) = \Gamma x(\tau = 0). \tag{47'}$$

Suppose that (49′) has a solution of the form

$$p(\tau) = R(\tau)x(\tau). \tag{51}$$

Then it follows immediately that

$$x(\tau) = A^{-1}(I + QR(\tau - 1))x(\tau - 1) \tag{52a}$$

$$R(\tau)x(\tau) = \left[\Xi A^{-1} + (A^T + \Xi A^{-1}Q)R(\tau - 1)\right]x(\tau - 1). \tag{52b}$$

Inserting (52a) in (52b) yields

$$R(\tau)A^{-1}(I + QR(\tau - 1)) = A^TR(\tau - 1) + \Xi A^{-1}(I + QR(\tau - 1))$$

or, equivalently,

$$R(\tau) = A^T\left[R(\tau-1)^{-1} + Q\right]^{-1} A + \Xi, \qquad \tau = 1, 2, \ldots, \beta \tag{53}$$

with $R(\tau = 0) = \Gamma$ in view of (47′). We notice that (53) is the well-known discrete Riccati equation (cf. (28) and (28′) in Section 2.2):

$$S(t-1) = A^T\left[S(t)^{-1} + Q\right]^{-1} A + \Xi, \qquad t = \beta, \beta-1, \ldots, 1 \tag{53′}$$

with $S(t = \beta) = \Gamma$. Hence the limit of $R(\tau)$ as $\tau \to \infty$ can be taken as the steady-state solution S in question.

Proceeding toward obtaining the limit of $R(\tau)$, we assume for the sake of brevity that all eigenvalues of matrix H are real and distinct. Let λ be an eigenvalue of H, and let x and p be n vector partitions of the associated eigenvector, i.e.,

$$\lambda\begin{pmatrix} x \\ p \end{pmatrix} = H\begin{pmatrix} x \\ p \end{pmatrix}.$$

Since (cf. Murata (1977, (31), Section 1.1))

$$(H^{-1})^T = \begin{pmatrix} A^T + \Xi A^{-1}Q & -\Xi A^{-1} \\ -A^{-1}Q & A^{-1} \end{pmatrix},$$

we have

$$\lambda\begin{pmatrix} -p \\ x \end{pmatrix} = (H^{-1})^T\begin{pmatrix} -p \\ x \end{pmatrix},$$

implying that λ is an eigenvalue of $(H^{-1})^T$. Consequently, λ is also an eigenvalue of H^{-1}, and in turn λ^{-1} is an eigenvalue of H. Thus the eigenvalues of H are such that the reciprocal of each eigenvalue is also an eigenvalue of H. If W is a nonsingular Jordan transformation matrix composed of linearly independent eigenvectors (cf. Murata (1977, Theorem 6, Section 3.2)), then

$$W^{-1}HW = \begin{pmatrix} \Lambda & 0 \\ 0 & \Lambda^{-1} \end{pmatrix} \tag{54}$$

where Λ is a diagonal matrix of the n eigenvalues outside the unit circle. Matrix W can be partitioned into four $n \times n$ matrices as follows:

$$W = \begin{pmatrix} W_{11} & W_{12} \\ W_{21} & W_{22} \end{pmatrix}. \tag{55}$$

Let $(z^T(\tau), q^T(\tau))^T$ be a new state vector defined by the transformation:

$$\begin{pmatrix} x(\tau) \\ p(\tau) \end{pmatrix} = \begin{pmatrix} W_{11} & W_{12} \\ W_{21} & W_{22} \end{pmatrix}\begin{pmatrix} z(\tau) \\ q(\tau) \end{pmatrix}. \tag{56}$$

Then we get from (49′), (54), and (56)

$$\begin{pmatrix} z(\tau) \\ q(\tau) \end{pmatrix} = \begin{pmatrix} \Lambda & 0 \\ 0 & \Lambda^{-1} \end{pmatrix} \begin{pmatrix} z(\tau-1) \\ q(\tau-1) \end{pmatrix}, \qquad \tau = 1, 2, \ldots, \beta$$

or by iterative substitutions

$$\begin{pmatrix} z(\tau) \\ q(\tau) \end{pmatrix} = \begin{pmatrix} \Lambda^{\tau} & 0 \\ 0 & \Lambda^{-\tau} \end{pmatrix} \begin{pmatrix} z(0) \\ q(0) \end{pmatrix},$$

which is equivalent to

$$\begin{pmatrix} z(0) \\ q(\tau) \end{pmatrix} = \begin{pmatrix} \Lambda^{-\tau} & 0 \\ 0 & \Lambda^{-\tau} \end{pmatrix} \begin{pmatrix} z(\tau) \\ q(0) \end{pmatrix}. \tag{57}$$

The boundary condition (47′) can be written in terms of $z(0)$ and $q(0)$ in view of (56) as

$$W_{21}z(0) + W_{22}q(0) = \Gamma\left[W_{11}z(0) + W_{12}q(0) \right],$$

from which it follows that

$$q(0) = Uz(0) \tag{58}$$

where

$$U \equiv -\left[W_{22} - \Gamma W_{12} \right]^{-1}\left[W_{21} - \Gamma W_{11} \right]. \tag{59}$$

From (57) and (58), we derive

$$q(t) = G_{\tau}z(\tau) \tag{60}$$

where

$$G_{\tau} \equiv \Lambda^{-\tau}U\Lambda^{-\tau}. \tag{61}$$

Substituting (60) into (56) yields

$$x(\tau) = \left[W_{11} + W_{12}G_{\tau} \right]z(\tau)$$

$$p(\tau) = \left[W_{21} + W_{22}G_{\tau} \right]z(\tau)$$

Connecting these relations with one another, we finally obtain the form

$$p(\tau) = R(\tau)x(\tau), \qquad \tau = 1, 2, \ldots, \beta \tag{51}$$

where

$$R(\tau) \equiv \left[W_{21} + W_{22}G_{\tau} \right]\left[W_{11} + W_{12}G_{\tau} \right]^{-1}. \tag{62}$$

$R(\tau)$ in (62) is nothing other than a solution to the Riccati difference equation (53′). Our present purpose is to find a nonrecursive solution for the limit of $R(\tau)$ as $\tau \to \infty$. Letting τ go to infinity, G_{τ} in (61) will approach zero matrix, and hence from (62) we have

$$S = \lim_{\tau \to \infty} R(\tau) = W_{21}W_{11}^{-1}. \tag{63}$$

In this computation we need only $(W_{11}^T, W_{21}^T)^T$, the first n eigenvectors associated with the eigenvalues of matrix H outside the unit circle. A numerical example of the computation is in order here.

EXAMPLE. Taking numerical data from an economic optimization problem in Section 6.5, we have the following relevant matrices (cf. (108), (90), (91) and (76) in Chapter 6):

$$A = \begin{bmatrix} -20 & 0 & -20 \\ -7.4 & -0.5 & -8.53 \\ 0 & 0 & 1 \end{bmatrix} \qquad \Phi = \begin{bmatrix} 0.5 & 0 & 0 \\ 0 & 1 & 0 \\ 0 & 0 & 1 \end{bmatrix}$$

$$B = \begin{bmatrix} 0 & -133.33 & 3333.33 \\ -1.67 & -48.66 & 1233.33 \\ 0 & 0 & 1 \end{bmatrix} \qquad \Xi = \begin{bmatrix} 1 & 0 & 0 \\ 0 & 60 & 0 \\ 0 & 0 & 2 \end{bmatrix}.$$

Then, matrix H in (50) is calculated from these data as

$$H = \begin{bmatrix} -0.05 & 0 & -1 & -559776.59 & -207112.50 & -167.67 \\ 0.74 & -2 & -2.26 & -7339.95 & -2727.95 & -2.26 \\ 0 & 0 & 1 & 3333.33 & 1233.33 & 1 \\ -0.05 & 0 & -1 & -559796.59 & -207119.90 & -167.67 \\ 44.40 & -120 & -135.60 & -440397.04 & -163677.55 & -135.35 \\ 0 & 0 & 2 & 6646.66 & 2458.13 & 3 \end{bmatrix}$$

of which the eigenvalues λ_i are found to be real and distinct, i.e.,

$$\lambda_1 = -722890.875, \qquad \lambda_2 = -583.227, \qquad \lambda_3 = 1.142,$$

$$\lambda_4 = 0.807, \qquad \lambda_5 = 0.013, \qquad \lambda_6 = 0.$$

The eigenvectors k_i associated with eigenvalues $\lambda_i (i = 1, 2, 3)$ outside the unit circle are shown next in column vectors, respectively:

$$[k_1 \quad k_2 \quad k_3] =$$

$$\left.\begin{bmatrix} -0.000021 & 0.060265 & -0.054231 \\ 0.000058 & -0.162752 & 0.073330 \\ 0.000066 & -0.182808 & 0.993003 \\ \cdots & \cdots & \cdots \\ 0.937750 & 0.561403 & 0.060621 \\ 0.347312 & -0.788205 & -0.001165 \\ 0.000283 & -0.004471 & 0.044163 \end{bmatrix} \begin{matrix} \left.\vphantom{\begin{matrix}a\\a\\a\end{matrix}}\right\} W_{11} \\ \\ \left.\vphantom{\begin{matrix}a\\a\\a\end{matrix}}\right\} W_{21} \end{matrix} \right. .$$

Of the above (6×3) matrix $[k_1, k_2, k_3]$, the upper half (3×3) matrix and the lower half (3×3) matrix correspond to W_{11} and W_{21} in (55), respectively. Thus S in (63) can be computed numerically using these values as

$$S = \begin{bmatrix} 1.305741 & -4.938547 & 0.000197 \\ -0.008557 & 4.853038 & -0.000279 \\ -0.003215 & 0.014727 & -0.000002 \end{bmatrix}.$$

2.4. Application to Control of a Dynamic Leontief System

We consider a dynamic quantity system of Leontief type:

$$x_t = Ax_t + B\{x_t - x_{t-1}\} + y_t + c \tag{64}$$

where A denotes the $n \times n$ nonnegative matrix whose (i, j)th component a_{ij} represents the quantity of good i consumed in the production of good j per unit of output, $B = [b_{ij}]$ denotes the corresponding capital coefficient matrix, $x_t = \{x_{1t}, \ldots, x_{nt}\}$ is the column vector of outputs in period t, and $c = \{c_1, \ldots, c_n\}$ and $y_t = \{y_{1t}, \ldots, y_{nt}\}$ are the column vectors of the constant part and the variable part of final demand, respectively. Equation (64) means that outputs are supplied to meet the total demand composed of current interindustry demand, investment (filling the gap between output capacity of the present period and that of the last), and final demand. (For the corresponding price system, refer to Murata (1977, Section 4.4).)

Solow (1959) treated final demand as a constant, but we add to it a variable final demand and treat the latter as an instrument to attain an asymptotic stability from the dynamic systems control viewpoint. Assuming that $[I - A]$ fulfills the Hawkins–Simon conditions (cf. Murata (1977, p. 52)), we have

$$M \equiv [I - A - B]^{-1} = \left[I - [I - A]^{-1}B\right]^{-1}[I - A]^{-1}, \tag{65}$$

and express system (64) in a state-space form

$$x_t = -MBx_{t-1} + My_t + Mc \tag{66}$$

with initial value x_0. System (66) is controllable in view of the nonsingularity of M. Let us adopt the quadratic loss function to be minimized: (denoting transportation by superscript T)

$$J = \sum_{t=1}^{\infty} \Big\{ \left(x_{t-1} - \alpha^{t-1}x^*\right)^T \Xi \left(x_{t-1} - \alpha^{t-1}x^*\right) + \left(y_t - \alpha^t y^*\right)^T \Gamma \left(y_t - \alpha^t y^*\right) \Big\}, \tag{67}$$

where Ξ and Γ are some constant symmetric positive definite matrices of appropriate dimensions, α denotes a growth factor $(= 1 + \text{growth rate } g)$, and $\alpha^t x^*$, $\alpha^t y^*$ are target values of x_t, y_t, respectively.

In some multisectoral turnpike problems, the objective is to maximize output value (see Tsukui (1966)) or utility (see Tsukui (1967)) over time, while we intend to minimize a sort of disutility in (67) over the time of the output and final-demand deviations from preassigned target growth paths. If parameter α is unity and if Ξ, Γ are identity matrices, our minimizing objective J reduces to

$$J' = \sum_{t=1}^{\infty} (\|x_{t-1} - x^*\| + \|y_t - y^*\|), \tag{67'}$$

which is close to the objective of Radner (1961).

By applying Theorem 17′ to our problem, we obtain the optimal control value of y_t and the associated output vector x_t for $t = 1, 2, \ldots$ as follows:

$$y_t = Gx_{t-1} + h_t - c^* \tag{68}$$

$$x_t = F^t x_0 + Mh_t + FMh_{t-1} + F^2Mh_{t-2} + \cdots + F^{t-1}Mh_1 - M(c^* - c), \tag{69}$$

where

$$G \equiv [M^TSM + \Gamma]^{-1} M^TSMB, \tag{70}$$

$$S \equiv B^TM^TSM[B - G] + \Xi = B^T\Gamma G + \Xi, \tag{71}$$

$$h_t \equiv [M^TSM + \Gamma]^{-1}\Big(M^T[I + \alpha F + \alpha^2F^2 + \ldots]^T \alpha^t Sx^* + \alpha^t \Gamma y^* + M^T[I + \alpha F + \alpha^2 F^2 + \ldots]^T \alpha^{t+1} G^T \Gamma y^*\Big), \tag{72}$$

$$F \equiv [f_{ij}] \equiv M[G - B]. \qquad (f_{ij} : \text{the } (i, j)\text{th element of } F) \tag{73}$$

$$c^* \equiv [M^TSM + \Gamma]^{-1} M^T[I - F^T]^{-1} SMc. \tag{74}$$

Premultiplying (70) by M and substituting for MG in (73) yields

$$F + MB = [M^TS + \Gamma M^{-1}]^{-1} M^TSMB$$

or, equivalently,

$$[M^TS + \Gamma M^{-1}][F + MB] = M^TSMB,$$

entailing

$$F = -[M^TS + \Gamma M^{-1}]^{-1} \Gamma B, \tag{73'}$$

$$G = B + M^{-1}F = \Big[I - [M^TSM + \Gamma]^{-1}\Gamma\Big]B, \tag{70'}$$

$$S = B^T\Gamma\Big[I - [M^TSM + \Gamma]^{-1}\Gamma\Big]B + \Xi. \tag{71'}$$

F is already known to be stable, i.e.,

$$\lim_{t\to\infty} F^t = 0. \tag{75}$$

(Refer to the proof of Theorem 15.) Our interest here is to find a sufficient condition for x_t in (69) to converge to some growth equilibrium characterized by α^t alone. We shall show that this can be achieved if each eigenvalue of αF is less than unity in modulus. In fact, if

$$-1 < g < \frac{1}{\max_j \sum_i |f_{ij}|} - 1 \tag{76}$$

is fulfilled, clearly every eigenvalue of αF is less than unity in modulus, and hence

$$I + \alpha F + \alpha^2 F^2 + \ldots = [I - \alpha F]^{-1}. \tag{77}$$

Therefore, x_t in (69) reduces to

$$\begin{aligned} x_t = {} & F^t x_0 + [\alpha^t I - F^t][I - \alpha^{-1} F]^{-1} M [M^T S M + \Gamma]^{-1} \\ & \times \left(M^T [I - \alpha F^T]^{-1} S x^* + \left[I + \alpha M^T [I - \alpha F^T]^{-1} G^T \right] \Gamma y^* \right) \\ & - M(c^* - c). \end{aligned} \tag{69'}$$

In view of (75), x_t in (69′) converges to a growth equilibrium characterized by α^t alone, as t goes to infinity. Thus we have the following.

Proposition 1. *Consider the problem of minimizing the criterion* (67) *for the system* (64). *If growth rate g is given to satisfy* (76), *the optimal control y_t regulated by* (68) *will stabilize the associated output vector x_t so that it converges to a growth equilibrium as t goes to infinity.*

By choosing appropriate positive definite matrices Ξ, Γ and nonnegative vectors x^*, y^*, obtaining nonnegative values of the optimal y_t, x_t for all t is possible.

As an illustration, we consider the one-good economy such that the quantity system (64) holds as a single equation, where coefficients A, B are regarded as positive scalars and set Ξ, Γ equal to unity in the criterion function (67). Then, utilizing (71′) and (70′), we calculate the relevant parameters in the optimal control as follows:

$$G = \frac{\sqrt{D^2 + 4B^2} - D}{2B} > 0 \tag{70''}$$

$$F = \frac{\sqrt{D^2 + 4B^2} - D - 2B^2}{2B(1 - A - B)} > 0, \tag{73''}$$

where $D \equiv 1 + (1 - A)(1 - A - 2B)$, and we assume $0 < 1 - A < B$. Finally, the growth equilibium value of output in period t is calculated as

$$x_t^* = \frac{4\alpha^t B^2 (1 - A - B)^2 \left[\left(\sqrt{D^2 + 4B^2} - D + 2 \right) x^* + 2(1 - A + gB) y^* \right]}{\left(\sqrt{D^2 + 4B^2} + D + 2B^2 \right) \left[2\alpha B(1 - A - B) + D + 2B^2 - \sqrt{D^2 + 4B^2} \right]^2} + \frac{c - c^*}{1 - A - B} \tag{69''}$$

which takes on a positive value, as will be shown, provided g, x^*, and y^* are all positive.

We show that $c^* - c > 0$, viz.,

$$(2MS + 1)^{-1} M (1 - F)^{-1} SM - 1 > 0.$$

Since $S = \left(\sqrt{D^2 + 4B^2} - D\right)/2 + 1$, we have

$$(2MS+1)^{-1}M(1-F)^{-1}SM$$

$$= \frac{2B(1-A-B)\left(\sqrt{D^2+4B^2} - D + 2\right)}{\left\{\left[\left(\sqrt{D^2+4B^2} - D + 2\right) + 2(1-A-B)^2\right] \times \left[2B(1-A-B) - \left(\sqrt{D^2+4B^2} - D - 2B^2\right)\right]\right\}}$$

$$= \frac{2B(1-A-B)\left(\sqrt{D^2+4B^2} - D + 2\right)}{2B(1-A-B)\left(\sqrt{D^2+4B^2} - D + 2\right) + 4B(1-A)(1-A-B)^2},$$

since $D = 1 + (1-A)(1-A-2B)$. The denominator in the extreme right-hand side of the expression reduces to $2B(1-A-B)(\sqrt{D^2+4B^2} + 1 + (1-A)^2)$. Thus

$$(2MS+1)^{-1}M(1-F)^{-1}SM - 1 = \frac{-2B(1-A-B)(1-A)}{\sqrt{D^2+4B^2} + 1 + (1-A)^2} > 0. \tag{78}$$

So far, we have considered equation systems for optimization problems, but we may consider inequality systems instead. Thus for our present application, we may consider

$$x_t \geqslant Ax_t + B\{x_t - x_{t-1}\} + y_t + c \tag{64'}$$

instead of (64), meaning that total output should not be smaller than total demand. Accordingly, equation (66) will change into an inequality. In the following section we try to obtain some control rules for such an inequality optimization problem.

2.5. Controller for a Dynamic Inequality System

We are concerned with a finite-horizon minimization problem with the same quadratic cost criterion as (16) subject to the linear discrete-time inequality system

$$x(t) \geqslant Ax(t-1) + Bv(t) + c(t) \tag{79}$$

for $t = 1, \ldots, \beta$ with given initial value $x(0) = x_0$. A distinct difference between the present problem and the one described in Section 2.2 lies in the inequality sign of system (79), all the notations being the same as for system (17).

Our approach to the inequality problem is not that of dynamic programming but quadratic programming. For convenience, we transform the problem into (80), as follows.

Minimize

$$\tilde{J} \equiv \tfrac{1}{2} z^T Q z \tag{80a}$$

subject to

$$Gz + b \leqslant 0, \tag{80b}$$

$$z \geqslant 0, \tag{80c}$$

where superscript T denotes transposition, and

$$z \equiv \begin{bmatrix} x(1) \\ \vdots \\ x(\beta-1) \\ x(\beta) \\ v(1) \\ \vdots \\ v(\beta) \end{bmatrix}, \qquad Q \equiv \begin{bmatrix} \Xi &&&&&& \\ & \ddots &&&&& \\ && \Xi &&&& \\ &&& \Gamma &&& \\ &&&& \Phi && \\ &&&&& \ddots & \\ &&&&&& \Phi \end{bmatrix} \tag{81}$$

$$G \equiv \begin{bmatrix} -I &&&&& B &&&& \\ A & -I &&&&& B &&& \\ & A & \ddots &&&&& \ddots && \\ && \ddots & \ddots &&&&& \ddots & \\ &&& A & -I &&&&& B \\ &&&& A & -I &&&& & B \end{bmatrix},$$

$$b \equiv \begin{bmatrix} c(1) + Ax_0 \\ c(2) \\ \vdots \\ \vdots \\ c(\beta) \end{bmatrix}. \tag{82}$$

Obviously, minimizing $\tilde{J}$ of (80a) is simply minimizing J of (16), and (80b) is equivalent to the set of inequalities (79) for $t = 1, \ldots, \beta$ with $x(0) = x_0$. Expression (80c) is an additional restriction specific to quadratic programming problem; viz., we restrict all the relevant variables (state and control variables) to nonnegative values. This restriction is meaningful in economics because most economic level variables assume nonnegative values. Note that matrix Q in (81) is symmetric and positive semidefinite, since Γ, Ξ, and Φ are assumed to be so (with Φ positive definite). Hence $\tilde{J}$ of (80a) is a convex function in z (cf. Murata (1977, Theorem 11, Section 7.2)).

Now we see that (80) is a typical quadratic programming problem. Thus by the Kuhn–Tucker theorem (Murata (1977, Theorem 25, Section 8.4)) we have the following necessary and sufficient condition for z to be an optimal solution to problem (80): a row $n\beta$ vector

$$p^T \equiv (p^T(1), \ldots, p^T(\beta)) \geqslant 0 \tag{83}$$

exists such that

$$p^T(Gz + b) = 0, \tag{84}$$

$$z^TQ + p^TG \geqslant 0, \tag{85}$$

$$(z^TQ + p^TG)z = 0. \tag{86}$$

(Cf. Murata (1977, Application 3, pp. 311–312).) Inequality (85) can be decomposed as

$$p(t) \leqslant A^Tp(t+1) + \Xi x(t) \qquad \text{for} \quad t = 1, \ldots, \beta - 1 \tag{85a}$$

$$p(\beta) \leqslant \Gamma x(\beta) \qquad \text{(transversality condition)} \tag{85b}$$

$$B^Tp(t) \geqslant -\Phi v(t) \qquad \text{for} \quad t = 1, \ldots, \beta. \tag{85c}$$

Assume that Φ has an inverse with all nonnegative entries. Then optimal control $v(t)$ must satisfy (85′c) in view of (85c).

$$v(t) \geqslant -\Phi^{-1}B^Tp(t) \qquad \text{for} \quad t = 1, \ldots, \beta. \tag{85′c}$$

Assuming further that

$$A^{-1} \geqslant 0, \qquad B \geqslant 0, \tag{*}$$

we get from (85a)

$$p(t+1) \geqslant (A^{-1})^T\{p(t) - \Xi x(t)\} \qquad \text{for} \quad t = 1, \ldots, \beta - 1. \tag{85′a}$$

Combining (85′a) with (85′c) yields for $t = 2, 3, \ldots, \beta$

$$v(t) \geqslant \Phi^{-1}B^T(A^{-1})^T\Xi x(t-1) - \Phi^{-1}B^T(A^{-1})^Tp(t-1). \tag{87}$$

As for $v(1)$, we have the following requirements in view of (79) and (85′c):

$$x(1) - Ax_0 - c(1) \geqslant Bv(1) \geqslant -B\Phi^{-1}B^Tp(1). \tag{88}$$

Given $x_0, x(1)$ and $p(1)$, we choose $v(1)$ to fulfill (88), and $v(2), \ldots, v(\beta)$ must be chosen to fulfill (87). At the same time $p(2), \ldots, p(\beta)$ should satisfy (85′a) and (85b). Note that, by virtue of (86), if some $x(t)$ is strictly positive, then the corresponding relation (85′a) holds with the equality sign.

Now we make a specification of $p(t)$ as

$$p(t) = S(t)x(t) \qquad \text{for} \quad t = 1, \ldots, \beta \tag{89}$$

with $S(\beta)=\Gamma$. Substituting this specification into (85a) and (85′c), we get

$$[S(t)-\Xi]x(t) \leqslant A^T S(t+1)p(t+1) \qquad \text{for} \quad t=1,\ldots,\beta-1, \tag{85″a}$$

and

$$v(t) \geqslant -\Phi^{-1}B^T S(t)x(t) \qquad \text{for} \quad t=1,\ldots,\beta. \tag{85″c}$$

Substitution of $v(t)$ from (85″c) into (79) yields, assuming (*),

$$[I+B\Phi^{-1}B^T S(t)]x(t) \geqslant Ax(t-1)+c(t) \qquad \text{for} \quad t=1,\ldots,\beta,$$

from which we obtain

$$x(t) \geqslant [I+B\Phi^{-1}B^T S(t)]^{-1}\{Ax(t-1)+c(t)\} \qquad \text{for} \quad t=1,\ldots,\beta \tag{90}$$

on the assumption that the relevant inverse is nonnegative:

$$[I+B\Phi^{-1}B^T S(t)]^{-1} = I - B[B^T S(t)B+\Phi]^{-1}B^T S(t) \geqslant 0. \tag{91}$$

If all the optimal $v(t)$'s are strictly positive, of if $S(t)$ are all nonnegative, relations (85c) and hence (85″c) hold with equality; and thus by (90) and (91), the optimal $v(t)$ is obtained as

$$v(t) \geqslant -\Phi^{-1}B^T S(t)\Big[I-B[B^T S(t)B+\Phi]^{-1}B^T S(t)\Big]\{Ax(t-1)+c(t)\},$$

or equivalently (cf. Murata (1977, p. 384)),

$$v(t) \geqslant -[B^T S(t)B+\Phi]^{-1}B^T S(t)\{Ax(t-1)+c(t)\} \tag{92}$$

for $t=1,\ldots,\beta$. This inequality feedback control rule (92) is comparable to the corresponding equality feedback control (26) in Theorem 14.

Concluding our quadratic programming approach, we should refer to Tamura (1975) for a similar attempt applied to a general distributed-lag equation system under some inequality constraints.

References

Aoki, M., *Optimal Control and System Theory in Dynamic Economic Analysis*. New York: North-Holland, 1976.

Bellman, R., *Dynamic Programming*. Princeton, NJ: Princeton Univ. Press, 1957.

Dorato, P., and Levis, A. H., "Optimal linear regulators: The discrete-time case," *IEEE Trans. Automatic Control*, vol. AC-16, pp. 613–620, 1971.

Hautus, M. L. J., "Controllability and observability conditions of linear autonomous systems," *Indagationes Mathematicae*, vol. 31, pp. 443–448, 1969.

———, "Stabilization, controllability and observability of linear autonomous systems," *Indagationes Mathematicae*, vol. 32, pp. 448–455, 1970.

Kalman, R. E., and Bertram, J. E., "Control system analysis and design via the "second method" of Lyapunov, II—Discrete-time systems," *J. Basic Engineering Trans. ASME*, vol. 82, pp. 394–400, 1960.

Kučera, V., "A contribution to matrix quadratic equations," *IEEE Trans. Automatic Control*, vol. AC-17, pp. 344–347, 1972.

Kwakernaak, H., and Sivan, R., *Linear Optimal Control Systems*. New York: Wiley-Interscience, 1972.

Mårtensson, K., "On the matrix Riccati equation," *Information Sciences*, vol. 3, pp. 17–49, 1971.

Murata, Y., *Mathematics for Stability and Optimization of Economic Systems*. New York: Academic Press, 1977.

Pappas, T., Laub, A. J., and Sandell, N. R., Jr., "On the numerical solution of the discrete-time algebraic Riccati equation," *IEEE Trans. Automatic Control*, vol. AC-25, pp. 631–641, 1980.

Radner, R., "Paths of economic growth that are optimal with regard only to final states: A turnpike theorem," *Review of Economic Studies*, vol. 28, pp. 98–104, 1961.

Solow, R. M., "Competitive valuation in a dynamic input-output system," *Econometrica*, vol. 27, pp. 30–53, 1959.

Tamura, H., "Decentralized optimization for distributed-lag models of discrete systems," *Automatica*, vol. 11, pp. 593–602, 1975.

Tinbergen, J., *On the Theory of Economic Policy*, 2nd ed. Amsterdam: North-Holland, 1952.

Tsukui, J., "Turnpike theorem in a generalized dynamic input-output system," *Econometrica*, vol. 34, pp. 396–407, 1966.

———, "The consumption and the output turnpike theorems in a von Neumann type of model—A finite term problem," *Review of Economic Studies*, vol. 34, pp. 85–93, 1967.

Vaughan, D. R., "A nonrecursive algebraic solution for the discrete Riccati equation," *IEEE Trans. Automatic Control*, vol. AC-15, pp. 597–599, 1970.

Wonham, W. M., "On pole assignment in multi-input controllable linear systems," *IEEE Trans. Automatic Control*, vol. AC-12, pp. 660–665, 1967.

CHAPTER 3

Observers for Linear Discrete-Time Systems

So far we have implicitly assumed that the state variables can all be observed in each period of time to establish control rules for linear discrete-time systems. However, in a real dynamic economy some of the variables will not be accessible in time. Section 3.2 of the present chapter is concerned with obtaining appropriate proxies (termed "observers") for the unobserved variables to be incorporated into optimal control laws and with the stability check of the resultant overall system. (Section 3.1 is devoted to preliminary propositions indispensable for constructing observers.) In Section 3.3 we show that the optimal control incorporating any observer incurs a cost rise, and hence in Section 3.4 we derive an observer that has the minimum cost performance. Finally, in Section 3.5, we examine the relationship between observer and controller and establish a separation principle for designing them with minimal overall associated cost.

3.1. Preliminaries to Discrete-Time Observers

If some state variables are not accessible (observable) in time, our optimal control rule must be modified so that unobserved state variables are replaced by their proxies. This problem is called the observer problem in control theory, and we consider it for the deterministic discrete-time system:

$$x(t) = Ax(t-1) + Bv(t), \tag{1}$$

where x is an n vector of state variables, v is an m vector of control variables, and A, B are constant matrices of appropriate dimensions.

In this section, we shall prove preliminary theorems indispensable to our observer problem. Assume hereafter, without loss of generality, that the first r elements of x, which will be denoted by an r vector y, are observed in system (1). Hence x is rewritten

$$x = \begin{pmatrix} y \\ w \end{pmatrix}, \tag{2}$$

where w represents an $(n - r)$ vector of unobserved state variables. We can express $y(t)$ as

$$y(t) = Cx(t), \tag{3}$$

where $C \equiv [I_r, 0]$ consists of an $r \times r$ identity matrix I_r and an $r \times (n - r)$ zero matrix. Accordingly, we partition A and B as

$$A = \begin{pmatrix} A_{11} & A_{12} \\ A_{21} & A_{22} \end{pmatrix} \quad \text{and} \quad B = \begin{pmatrix} B_{11} \\ B_{22} \end{pmatrix}, \tag{4}$$

and system (1) is rewritten

$$y(t) = A_{11}y(t-1) + A_{12}w(t-1) + B_{11}v(t) \tag{5a}$$

$$w(t) = A_{21}y(t-1) + A_{22}w(t-1) + B_{22}v(t). \tag{5b}$$

The following is a theorem concerning observability of this system.

Theorem 1 (Gopinath, 1971). *Let (A, C) be observable for the system consisting of* (1) *and* (3) *with $C = [I_r, 0]$. Then (A_{22}, A_{12}) is observable for system* (5).

PROOF.

$$n = rk \begin{bmatrix} C \\ CA \\ CA^2 \\ \vdots \\ CA^{n-1} \end{bmatrix} = rk \begin{bmatrix} I_r & 0 \\ A_{11} & A_{12} \\ A_{11}^2 + A_{12}A_{21}, & A_{11}A_{12} + A_{12}A_{22} \\ \vdots & \vdots \end{bmatrix}. \tag{6}$$

Since rank of a matrix is unaltered by adding to any row a linear combination of the other rows, we get from (6)

$$n = rk \begin{bmatrix} I_r & 0 \\ A_{11} & A_{12} \\ \vdots & A_{12}A_{22} \\ & A_{12}A_{22}^2 \\ & \vdots \\ & A_{12}A_{22}^{n-1} \end{bmatrix}. \tag{7}$$

For example, the third row block of the matrix in (7) is obtained by adding to the third row block of the matrix in (6)

$$-A_{11}^2 \cdot (\text{its first row block}) - A_{11} \cdot (\text{its second row block}).$$

From (7) we know that, regardless of what the first column block of the matrix in (7) may be, its second column block has rank $n - r$. In view of the Cayley–Hamilton theorem, therefore, we have

$$rk \begin{bmatrix} A_{12} \\ A_{12}A_{22} \\ A_{12}A_{22}^2 \\ \vdots \\ A_{12}A_{22}^{n-r-1} \end{bmatrix} = n - r. \qquad \square$$

In order to prove an important theorem to follow, we need a lemma.

Lemma 1. *Let A be an $n \times n$ real matrix, and assume $\lambda I - A$ is nonsingular. Then*

$$[\lambda I - A]^{-1} = \sum_{j=1}^{n} \rho_j(\lambda) A^{j-1},$$

where $\rho_j(\lambda)$ is some rational function of λ.

PROOF. Let A be an $n \times n$ real matrix.

$$\operatorname{adj}[\lambda I - A] = B_{n-1}\lambda^{n-1} + B_{n-2}\lambda^{n-2} + \cdots + B_2\lambda^2 + B_1\lambda + B_0,$$

where B_i $(i = 0, 1, \ldots, n-1)$ are matrices not containing λ. Since

$$[\lambda I - A] \cdot \operatorname{adj}[\lambda I - A] = |\lambda I - A| I$$

and since

$$|\lambda I - A| = \lambda^n + a_{n-1}\lambda^{n-1} + a_{n-2}\lambda^{n-2} + \cdots + a_2\lambda^2 + a_1\lambda + (-1)^n|A|,$$

we have

$$[\lambda I - A][B_{n-1}\lambda^{n-1} + B_{n-2}\lambda^{n-2} + B_{n-3}\lambda^{n-3} + \cdots + B_2\lambda^2 + B_1\lambda + B_0]$$
$$= \lambda^n I + a_{n-1}\lambda^{n-1} I + a_{n-2}\lambda^{n-2} I + \cdots + a_2\lambda^2 I + a_1\lambda I + (-1)^n|A|I.$$

Comparing both sides term by term, we know that the coefficients of $\lambda^n, \lambda^{n-1}, \lambda^{n-2}, \ldots, \lambda^2, \lambda$, and the constant term have the following relations:

$$B_{n-1} = I,$$
$$B_{n-2} - AB_{n-1} = a_{n-1}I \Rightarrow B_{n-2} = a_{n-1}I + A,$$
$$B_{n-3} - AB_{n-2} = a_{n-2}I \Rightarrow B_{n-3} = a_{n-2}I + a_{n-1}A + A^2,$$
$$\vdots$$
$$B_1 - AB_2 = a_2 I \Rightarrow B_1 = a_2 I + a_3 A + \cdots + A^{n-2},$$
$$B_0 - AB_1 = a_1 I \Rightarrow B_0 = a_1 I + a_2 A + \cdots + A^{n-1}.$$

Then consider

$$[\lambda I - A]^{-1} = \mathrm{adj}[\lambda I - A]/|\lambda I - A|. \qquad \square$$

Theorem 2 (Luenberger, 1971). *Let A and C be $n \times n$ and $r \times n$ real matrices, respectively. The assertion that, by the suitable choice of an $n \times r$ real matrix G, the set of eigenvalues of $A - GC$ can be made to correspond to any set of n distinct real scalars different from eigenvalues of A is true if and only if (A, C) is observable.*

PROOF.

Necessity (Cf. Wonham (1967).) Let $\lambda_1, \lambda_2, \ldots, \lambda_n$ be any distinct real scalars different from eigenvalues of matrix A, so $|\lambda_i I - A| \neq 0$ for each λ_i $(i = 1, \ldots, n)$. By our assertion, n row vectors p_i $(i = 1, \ldots, n)$ and an $n \times r$ matric G exist such that

$$p_i[\lambda_i I - A + GC] = 0, \qquad i = 1, \ldots, n;$$

and by the assumption that λ_i is different from any eigenvalue of A,

$$\begin{aligned} p_i &= -p_i GC[\lambda_i I - A]^{-1} \\ &= p_i \sum_{j=1}^{n} \rho_j(\lambda_i) GCA^{j-1}, \qquad i = 1, \ldots, n \end{aligned}$$

for some rational functions $\rho_j(\lambda_i)$ of λ_i $(j = 1, \ldots, n)$ by virtue of Lemma 1 above. We can also rewrite this set of equations in matrix form:

$$P = NR, \tag{8}$$

where

$$P \equiv \begin{bmatrix} p_1 \\ \vdots \\ p_n \end{bmatrix}, \qquad N \equiv \begin{bmatrix} p_1\rho_1(\lambda_1)G, & p_1\rho_2(\lambda_1)G, \ldots, & p_1\rho_n(\lambda_1)G \\ \vdots & \vdots & \vdots \\ p_n\rho_1(\lambda_n)G, & p_n\rho_2(\lambda_n)G, \ldots, & p_n\rho_n(\lambda_n)G \end{bmatrix},$$

$$R \equiv \begin{bmatrix} C \\ CA \\ \vdots \\ CA^{n-1} \end{bmatrix}.$$

Since $p_1, \ldots, p_n$ are linearly independent in view of the distinctness of λ_i $(i = 1, \ldots, n)$, we get

$$rk(P) = n. \tag{*}$$

On the other hand,

$$rk(P) \leqslant \min\{rk(N), rk(R)\} \leqslant n. \tag{**}$$

Expressions (*) and (**) imply that $rk(R) = n$, viz., (A, C) is observable.

Sufficiency (Cf. Gopinath (1971).) Let $\lambda_1, \ldots, \lambda_n$ be n distinct real scalars. Then constants $\gamma_1, \ldots, \gamma_n$ exist satisfying

$$\begin{bmatrix} \lambda_1^n \\ \vdots \\ \lambda_n^n \end{bmatrix} = - \begin{bmatrix} \lambda_1^{n-1} & \lambda_1^{n-2} \ldots \lambda_1 & 1 \\ \vdots & & \vdots \\ \lambda_n^{n-1} & \lambda_n^{n-2} \ldots \lambda_n & 1 \end{bmatrix} \begin{bmatrix} \gamma_1 \\ \vdots \\ \gamma_n \end{bmatrix}, \tag{9}$$

where we note that the determinant of the matrix on the right-hand side of a Vandermonde determinant equal to $\Pi_{k>i}(\lambda_k - \lambda_i)$. In other words, there a real matrix M corresponds to the given λ_i $(i = 1, \ldots, n)$ such that

$$|\lambda I - M| = \lambda^n + \sum_{j=1}^{n} \gamma_j \lambda^{n-j} \tag{10}$$

with λ_i as one of its eigenvalues.

Let G be an $n \times r$ real matrix of rank one. Then $GC(\lambda I - A)^{-1}$ has rank one since

$$rk\big[GC(\lambda I - A)^{-1}\big] = rk(GC) \leqslant \min\{rk(G), rk(C)\} = 1.$$

Thus

$$|I + GC(\lambda I - A)^{-1}| = 1 + \mathrm{tr}\big(GC(\lambda I - A)^{-1}\big) \tag{11}$$

since, for a square matrix S, we have (cf. Murata (1977, (10), Section 1.1))

$$|I + S| = |I| + \mathrm{tr}(S) + \begin{matrix}\text{(the sum of determinants involving} \\ \text{no-less-than-two columns of } S\text{)},\end{matrix}$$

of which the third term vanishes if S has rank one.

For a real square matrix A,

$$(\lambda I - A)^{-1} = \frac{1}{\lambda}\left(I - \frac{1}{\lambda}A\right)^{-1} = \sum_{i=0}^{\infty} A^i \lambda^{-(i+1)} \tag{12}$$

for λ different from an eigenvalue of A, in view of the corollary to Theorem 7 in Murata (1977, Section 3.3). From (11) and (12)

$$\begin{aligned} |\lambda I - A + GC| &= |\lambda I - A|\,|I + GC(\lambda I - A)^{-1}| \\ &= |\lambda I - A| + |\lambda I - A|\,\mathrm{tr}\left(GC\sum_{i=0}^{\infty} A^i \lambda^{-(i+1)}\right) \end{aligned} \tag{13}$$

for λ different from an eigenvalue of A. We consider

$$|\lambda I - A| = \lambda^n + \sum_{j=1}^{n} a_j \lambda^{n-j}$$

$$A^n = -\sum_{i=1}^{n} a_i A^{n-i},$$

and compare (13) term by term with the following:

$$|\lambda I - A + GC| \equiv \lambda^n + \sum_{j=1}^{n} \gamma_j \lambda^{n-j}. \tag{10'}$$

Then

$$\sum_{j=1}^{n}\gamma_j\lambda^{n-j}=\sum_{j=1}^{n}a_j\lambda^{n-j}+\left(\lambda^n+\sum_{i=1}^{n}a_i\lambda^{n-i}\right)$$
$$\times\operatorname{tr}\left[GC\left(\sum_{i=1}^{n}A^{i-1}\lambda^{-i}-\sum_{i=1}^{n}a_iA^{n-i}\lambda^{-n-1}+\sum_{i=n+1}^{\infty}A^{i}\lambda^{-(i+1)}\right)\right]$$
$$=\sum_{j=1}^{n}\lambda^{n-j}\left[a_j+\operatorname{tr}(GCA^{j-1})\right]$$
$$+\sum_{i=1}^{n}a_i\lambda^{n-i}\left[\sum_{i=1}^{n}\lambda^{-i}\operatorname{tr}(GCA^{i-1})+\dots\right], \tag{14}$$

from which we have

$$\begin{aligned}
&\gamma_j=0 \qquad \text{for} \quad j<0\\
&\gamma_1=a_1+\operatorname{tr}(GC)\\
&\gamma_2=a_2+a_1\operatorname{tr}(GC)+\operatorname{tr}(GCA)\\
&\vdots\\
&\gamma_n=a_n+a_{n-1}\operatorname{tr}(GC)+a_{n-2}\operatorname{tr}(GCA)+\cdots+\operatorname{tr}(GCA^{n-1}) \qquad (15)\\
&\gamma_j=0 \qquad \text{for} \quad j>n.
\end{aligned}$$

This proves that if G of rank one exists such that (10′) is fulfilled, then G satisfies (15), and that (10′) holds for any G of rank one satisfying (15).

Last, we shall show that at least one matrix G of rank one exists satisfying (15) if (A,C) is observable. Defining

$$\gamma\equiv\begin{bmatrix}\gamma_1\\ \vdots\\ \gamma_n\end{bmatrix},\qquad a\equiv\begin{bmatrix}a_1\\ \vdots\\ a_n\end{bmatrix},\qquad D\equiv\begin{bmatrix}1 & 0 & \cdots & \cdots & 0\\ a_1 & 1 & & & 0\\ a_2 & a_1 & \ddots & & \\ \vdots & & & \ddots & \\ a_{n-1} & a_{n-2} & \cdots & a_1 & 1\end{bmatrix},$$

we rewrite the set of n equations of γ_i $(i=1,\dots,n)$ in (15) as

$$\gamma=a+D\begin{bmatrix}\operatorname{tr}(GC)\\ \operatorname{tr}(GCA)\\ \vdots\\ \operatorname{tr}(GCA^{n-1})\end{bmatrix}. \tag{15′}$$

Letting $G=gh$, where g and h are $n\times 1$ and $1\times r$, respectively, we get

$$\operatorname{tr}(GCA^{i})=\operatorname{tr}(hCA^{i}g)=hCA^{i}g \tag{16}$$

for $i=0,1,\dots$. Substituting (16) in (15′) yields

$$\hat{h}_nRg=D^{-1}(\gamma-a), \tag{17}$$

where R is that in (8) and

$$\hat{h}_n \equiv \begin{bmatrix} h & & & \\ & h & & \\ & & \ddots & \\ & & & h \end{bmatrix}, \qquad (n \times rn).$$

Equation (17) has a unique solution for g, provided the product of matrices on the left-hand side of (17) has full rank. This will be fulfilled if (A, C) is observable, i.e., R has rank n. □

Note that in Theorem 2 above the $r \times n$ matrix C need not be of the form $[I_r, 0]$. However, if matrix C is of full rank equal to r, and $C = [C_1, C_2]$ where the $r \times r$ matrix C_1 is nonsingular, we can convert C into $C^* = [I_r, 0]$ by postmultiplying C by U^{-1} where

$$U \equiv \begin{pmatrix} C_1 & C_2 \\ 0 & I_{n-r} \end{pmatrix}$$

and hence

$$U^{-1} = \begin{pmatrix} C_1^{-1} & -C_1^{-1}C_2 \\ 0 & I_{n-r} \end{pmatrix}.$$

By applying the same conversion to

$$x(t) = Ax(t-1) + Bv(t) \tag{1}$$

$$y(t) = Cx(t), \tag{3}$$

we have

$$x^*(t) = A^*x^*(t-1) + B^*v(t) \tag{1'}$$

$$y(t) = C^*x^*(t) \tag{3'}$$

where

$$x^* \equiv Ux$$
$$A^* \equiv UAU^{-1}$$
$$B^* \equiv UB$$
$$C^* \equiv CU^{-1} = [I_r, 0].$$

Thus our discussion may address the system (1), (3) with $C = [I_r, 0]$ without loss of generality.

3.2. Luenberger Observers for Discrete-Time Systems

We are almost in a position to construct a proxy, called an observer, for unobserved state-variable vector w in (2). For this purpose, another lemma will be useful.

Lemma 2 (Luenberger, 1964). *Consider a discrete-time system composed of* (1) *and* (3), *and its associated order-*$(n-r)$ *system*:

$$z(t) = Fz(t-1) + Hy(t-1) + TBv(t), \tag{18}$$

where T *is an* $(n-r) \times n$ *transformation matrix satisfying*

$$TA - FT = HC. \tag{19}$$

Then $z(t)$ *can be expressed as*

$$z(t) = Tx(t) + F^t(z(0) - Tx(0)). \tag{20}$$

Proof. From (1), (3), and (18) it follows that

$$z(t) - Tx(t) = Fz(t-1) + HCx(t-1) - TAx(t-1),$$

into which we substitute (19) for HC, obtaining

$$z(t) - Tx(t) = F(z(t-1) - Tx(t-1)).$$

From this equation, we get (20) by iterative substitutions. □

Now we come to the main theorem on observers.

Theorem 3 (Luenberger, 1964). *Consider an order-*n *discrete-time system* (1), (3) *with* $C = [I_r, 0]$, *and assume that* (A, C) *is observable and that the last* $n-r$ *components of state vector* x *are unobserved variables, denoted* w *as in* (2). *Then an observer of order* $n-r$ *for the* w *can be constructed so that the involved matrix corresponds to an arbitrary set of distinct real eigenvalues, each of which has modulus less than unity and different from any eigenvalue of* A_{22} *in* (4).

Proof. Consider the partitions in (4) and define

$$T = [-G, I_{n-r}] \tag{21}$$

$$F = A_{22} - GA_{12} \tag{22}$$

for some $(n-r) \times r$ matrix G. Then

$$TB = B_{22} - GB_{11}. \tag{23}$$

H in (19) now becomes

$$H = A_{21} - GA_{11} + (A_{22} - GA_{12})G. \tag{24}$$

Thus (18) is expressed as

$$z(t) = [A_{22} - GA_{12}]z(t-1) + [A_{21} - GA_{11} + (A_{22} - GA_{12})G]y(t-1) + [B_{22} - GB_{11}]v(t). \tag{25}$$

Given the initial value $z(0)$ as

$$z(0) = w(0) - Gy(0), \tag{26}$$

we have

$$w(t) = z(t) + Gy(t) \equiv \hat{w}(t) \qquad \text{for} \quad t \geqslant 0, \tag{27}$$

since in view of (20)

$$w(t) = z(t) + Gy(t) + [A_{22} - GA_{12}]^t (w(0) - z(0) - Gy(0)). \quad (27')$$

Even if (26) does not hold, $w(t)$ may be approximated by $\hat{w}(t)$, an observer for $w(t)$, for large t provided $A_{22} - GA_{12}$ is a stable matrix, i.e.,

$$(A_{22} - GA_{12})^t \to 0 \qquad \text{as} \quad t \to \infty. \quad (28)$$

This is equivalent to saying that $A_{22} - GA_{12}$ has each of its eigenvalues less than unity in modulus.

By Theorem 1, if (A, C) is observable, so is (A_{22}, A_{12}). Then, by Theorem 2, the set of eigenvalues of $A_{22} - GA_{12}$ can be made to correspond to an arbitrary set of $n - r$ distinct real scalars different from eigenvalues of A_{22} by suitable choice of an $(n - r) \times r$ real matrix G. □

A remaining task is to choose an $(n - r) \times r$ matrix G such that the above requirements are fulfilled. The procedure for computing the matrix G is as follows.

Step 1. *Specify a set of s $(= n - r)$ distinct real scalars $\lambda_1, \ldots, \lambda_s$, having modulus less than unity and different from eigenvalues of A_{22}.*

Step 2. *Determine $\gamma_1, \ldots, \gamma_s$ as a unique solution of*

$$\begin{bmatrix} \lambda_1^s \\ \vdots \\ \lambda_s^s \end{bmatrix} = -\begin{bmatrix} \lambda_1^{s-1} & \lambda_1^{s-2} \ldots \lambda_1 & 1 \\ \vdots & & \vdots \\ \lambda_s^{s-1} & \lambda_s^{s-2} \ldots \lambda_s & 1 \end{bmatrix} \begin{bmatrix} \gamma_1 \\ \vdots \\ \gamma_s \end{bmatrix}. \quad (9')$$

Step 3. *Calculate the coefficients a_i $(i = 1, \ldots, s)$ in the characteristic polynomial*

$$|\lambda I - A_{22}| = \lambda^s + \sum_{j=1}^{s} a_j \lambda^{s-j},$$

viz.,

$$a_1 = -\operatorname{tr}(A_{22}), \ldots, \qquad a_s = (-1)^s |A_{22}|.$$

Step 4. *Define*

$$\tilde{\gamma} \equiv \begin{bmatrix} \gamma_1 \\ \vdots \\ \gamma_s \end{bmatrix}, \qquad \tilde{a} \equiv \begin{bmatrix} a_1 \\ \vdots \\ a_s \end{bmatrix}, \qquad \tilde{D} \equiv \begin{bmatrix} 1 & 0 & \cdots & \cdots & 0 \\ a_1 & 1 & & & 0 \\ a_2 & a_1 & \ddots & & \\ \vdots & & & \ddots & \\ a_{s-1} & a_{s-2} & \cdots & a_1 & 1 \end{bmatrix}.$$

Step 5. *Specify a row vector h of dimension r, and define $s \times rs$ matrix $\hat{h}_s$ as*

$$\hat{h}_s \equiv \begin{bmatrix} h & & & \\ & h & & \\ & & \ddots & \\ & & & h \end{bmatrix}, \quad \textit{and} \quad \tilde{A} \equiv \begin{bmatrix} A_{12} \\ A_{12}A_{22} \\ \vdots \\ A_{12}A_{22}^{s-1} \end{bmatrix}.$$

Step 6. *Compute column s-vector g by*

$$g = [\tilde{D}\hat{h}_s\tilde{A}]^{-1}(\tilde{\gamma} - \tilde{a}). \tag{17'}$$

Step 7. *Define G, $s \times r$ matrix, as the product of g in* (17′) *and h in Step* 5:

$$G \equiv gh. \tag{29}$$

Having obtained the matrix G, we may use $\hat{w}(t)$ in (27) as an observer for $w(t)$, with $z(t)$ substituted from (25), viz.,

$$\hat{w}(t) \equiv z(t) + Gy(t), \tag{30}$$

where

$$z(t) = [A_{22} - GA_{12}]z(t-1) + [A_{21} - GA_{11} + (A_{22} - GA_{12})G] \times y(t-1) + [B_{22} - GB_{11}]v(t) \tag{25}$$

with a given initial value $z(0)$. $\hat{w}(t)$ in (30) is termed a *Luenberger* (*minimal-order*) *observer* for the original system (1), (3) since the z has order $n - r$, the same order as unobserved state-variable vector w.

Our Luenberger observer $\hat{w}$ for unobserved state vector w is schematized in a flow chart (Fig. 1), where we see how z in (18) connects $\hat{w}$ with the original state-space system (1), (3).

Here we check the stability property of the whole state variables involving observers. Corresponding to (30), define an observer for the whole state variables as

$$\hat{x}(t) \equiv \begin{pmatrix} y(t) \\ \hat{w}(t) \end{pmatrix} = \begin{pmatrix} I_r & 0 \\ G & I_{n-r} \end{pmatrix} \begin{pmatrix} y(t) \\ z(t) \end{pmatrix}, \tag{31}$$

which may be rewritten as follows with (3) taken into account:

$$\hat{x}(t) = QCx(t) + \tilde{E}z(t), \tag{31'}$$

where

$$Q \equiv \begin{pmatrix} I_r \\ G \end{pmatrix}, \qquad \tilde{E} \equiv \begin{pmatrix} 0 \\ I_{n-r} \end{pmatrix}.$$

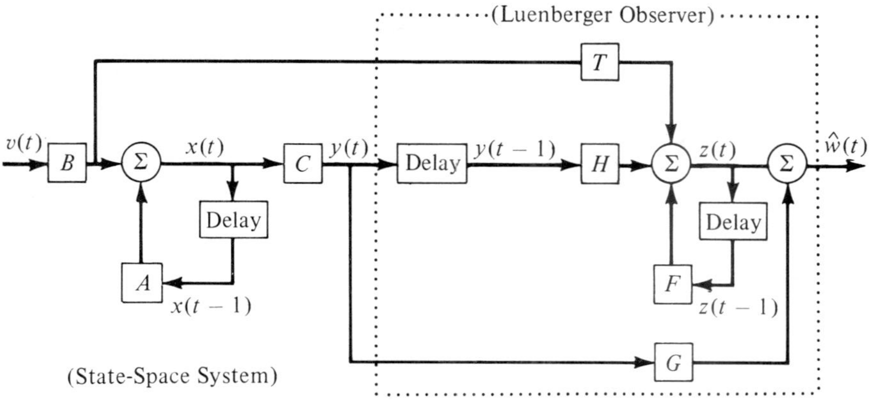

Figure 1. Luenberger Observer Connected with State-Space System
($\sum$ is summation, and Delay means one period delay.)

Thus the optimal control incorporating our observer becomes

$$v(t) = -K\hat{x}(t-1) \equiv \hat{v}(t), \tag{32}$$

and the associated behavior of state variables is governed by

$$\begin{aligned} x(t) &= Ax(t-1) - BK\hat{x}(t-1) \\ &= Ax(t-1) - BK(QCx(t-1) + \tilde{E}z(t-1)) \\ &= [A - BKQC]x(t-1) - BK\tilde{E}z(t-1), \end{aligned} \tag{33}$$

where K is the gain matrix defined by ($27^{\dagger}$) in Section 2.2. We shall show that the incorporation of our observers does not affect the stability property of the associated behavior of whole state variables. (For continuous-time cases, see Luenberger (1966).)

We must consider a composite system consisting of (33) and the behavior of $z(t)$ given by (25) incorporated with $v(t)$ of (32). (Aoki and Huddle (1967) will be useful for the subsequent development.) Substituting (3), (32), and (31′) into (25) with matrix notations (21)–(23) taken into consideration yields

$$\begin{aligned} z(t) &= Fz(t-1) + TAQCx(t-1) - TBK(QCx(t-1) + \tilde{E}z(t-1)) \\ &= T[A - BK]QCx(t-1) + [F - TBK\tilde{E}]z(t-1). \end{aligned} \tag{34}$$

The stability property of the system composed of (33) and (34) is characterized by its coefficient matrix, but the matrix is rather complicated. We therefore introduce a new variable vector e defined as

$$e(t) \equiv z(t) - Tx(t) = F^{t}(z(0) - Tx(0)). \tag{35}$$

Then vectors z in (33) and (34) are replaced by

$$z(t) = e(t) + Tx(t). \tag{20$'$}$$

This yields

$$\begin{aligned} x(t) &= [A - BKQC]x(t-1) - BK\tilde{E}(e(t-1) + Tx(t-1)) \\ &= [A - BK]x(t-1) - BK\tilde{E}e(t-1) \end{aligned} \tag{33$'$}$$

since $QC + \tilde{E}T = I$, and

$$\begin{aligned} e(t) = &-Tx(t) + T[A - BK]QCx(t-1) \\ &+ [F - TBK\tilde{E}](e(t-1) + Tx(t-1)) \\ = &(-T[A - BK] + T[A - BK]QC + [F - TBK\tilde{E}]T)x(t-1) \\ &+ (TBK\tilde{E} + F - TBK\tilde{E})e(t-1) \\ = &Fe(t-1) \end{aligned} \tag{34$'$}$$

since $F = TA\tilde{E}$. The composite system consisting of (33′) and (34′)

$$\begin{pmatrix} x(t) \\ e(t) \end{pmatrix} = \begin{pmatrix} A - BK & -BK\tilde{E} \\ 0 & F \end{pmatrix} \begin{pmatrix} x(t-1) \\ e(t-1) \end{pmatrix} \tag{36}$$

inherits the same stability property as the system composed of (33) and (34). The eigenvalues of the system (36) are those of $A - BK$ and of $F \equiv A_{22} - GA_{12}$. Therefore, when G is selected so that F has all its eigenvalues less than unity in modulus, the stability property of system (36), and hence that of the optimal behavior of state variables, is not affected by incorporation of our observer into optimal feedback control rule.

Concluding this section, we suggest our Luenberger observer can be extended to any distributed-lag system:

$$x(t) = \sum_{i=1}^{k} A_i x(t-i) + \sum_{j=0}^{h} B_j v(t-j) \tag{37}$$

with output equation

$$y(t) = Cx(t), \tag{38}$$

where $x(t)$, $v(t)$, and $y(t)$ are state n vector, control m vector, and output r vector, respectively, in period t; and A_i, B_j, C are constant matrices of appropriate dimensions with $rk(C) = r < n$. Letting $y(t)$ consist of the first r components of $x(t)$, we have

$$C = [I_r, 0] \tag{39}$$

and the rest of $x(t)$ is an unobserved state-variable vector of order $n - r$.

System (37) can be rewritten

$$\tilde{x}(t) = \tilde{A}\tilde{x}(t-1) + \tilde{B}v(t), \tag{37'}$$

where

$$\tilde{x}(t) \equiv \begin{bmatrix} v(t-h+1) \\ \vdots \\ v(t-1) \\ v(t) \\ x(t-k+1) \\ \vdots \\ x(t-1) \\ x(t) \end{bmatrix}, \quad \tilde{A} \equiv \begin{bmatrix} 0 & & I_m & & 0 & 0 & \cdots & \cdots & \cdots & 0 \\ \vdots & \ddots & & \ddots & & \vdots & & & & \vdots \\ \vdots & & 0 & & I_m & 0 & \cdots & \cdots & \cdots & 0 \\ 0 & \cdots & 0 & & 0 & 0 & \cdots & \cdots & \cdots & 0 \\ 0 & \cdots & \cdots & \cdots & 0 & 0 & & I_n & & 0 \\ \vdots & & & & \vdots & \vdots & \ddots & & \ddots & \\ 0 & \cdots & \cdots & \cdots & 0 & 0 & \cdots & 0 & & I_n \\ B_h & \cdots & \cdots & \cdots & B_1 & A_k & \cdots & A_2 & & A_1 \end{bmatrix}, \quad \tilde{B} \equiv \begin{bmatrix} 0 \\ \vdots \\ 0 \\ I_m \\ 0 \\ \vdots \\ 0 \\ B_0 \end{bmatrix}.$$

Accordingly, (38) is expressed as

$$\tilde{y}(t) = \tilde{C}\tilde{x}(t), \tag{38'}$$

with

$$\tilde{C} = [I_{\tilde{r}}, 0], \tag{39'}$$

where $\tilde{r} \equiv hm + kn - n + r$, and

$$\tilde{y}(t) \equiv \begin{bmatrix} v(t-h+1) \\ \vdots \\ v(t-1) \\ v(t) \\ x(t-k+1) \\ \vdots \\ x(t-1) \\ y(t) \end{bmatrix}.$$

Note that the dimensions of $\tilde{x}(t)$ and $\tilde{y}(t)$ are $\tilde{n}$ ($\equiv hm + kn$) and $\tilde{r}$ ($\equiv \tilde{n} - n + r$), respectively. Clearly, $rk(\tilde{C}) = \tilde{r} < \tilde{n}$. Thus the transformed state-space representation (37′), (38′) is of the same form as system (1), (3). Hence we can follow a procedure similar to our previous steps for obtaining a Luenberger observer for unobserved $n - r$ variables in $x(t)$.

However, dimensionality problems may occur with observer design for large-scale systems such as (37′), (38′). Refer to Arbel and Tse (1979) for an approach to reducing the computational requirements in observer design for large-scale linear discrete-time systems.

3.3. Cost Performance of Optimal Control Incorporating Observers

In this section we verify a cost rise due to the introduction of a Luenberger observer. It is known that if state variables were all perfectly observed in each time, we would adopt the standard optimal feedback control rule (cf. Theorem 15 in Section 2.2)

$$v(t+1) = -Kx_*(t), \tag{40}$$

where x_* denotes the associated state vector governed by

$$x_*(t) = [A - BK]x_*(t-1) \tag{41}$$

with $x_*(0) = x(0)$, ensuring a minimum of quadratic cost function:

$$J \equiv \sum_{t=0}^{\infty} \{x_*^T(t)\Xi x_*(t) + v^T(t+1)\Phi v(t+1)\}. \tag{42}$$

When state variables are not perfectly observed as we assume, we apply $\hat{x}$ instead of x to feedback control rule as in (32), obtaining an associated cost rise ΔJ above the minimal value J^* of J in (42):

$$\Delta J \equiv \hat{J} - J^*, \tag{42'a}$$

$$J^* \equiv x^T(0)Sx(0) \qquad \text{with} \quad S = A^TS[A - BK] + \Xi, \tag{42'b}$$

$$\hat{J} \equiv \sum_{t=0}^{\infty} \{x^T(t)\Xi x(t) + \hat{v}^T(t+1)\Phi\hat{v}(t+1)\}. \tag{42'c}$$

The vector $e(t)$ defined in (35), an estimation error, is rewritten

$$e(t) = \hat{w}(t) - w(t) \tag{35'}$$

since $Tx(t) = -Gy(t) + w(t)$ and since $\hat{w}(t) = z(t) + Gy(t)$. Then $\hat{x}(t)$ in (31) can be expressed as

$$\hat{x}(t) = x(t) + \tilde{E}e(t), \tag{43}$$

where $\tilde{E}$ is that appearing in (31′). Equation (43) means that measurement of the state vector involves estimation error $\tilde{E}e(t)$. $x(t)$ itself will differ from the $x_*(t)$ mentioned above by $e_*(t)$, say:

$$x(t) = x_*(t) + e_*(t) \tag{44}$$

with $e_*(0) = 0$ since $x_*(0) = x(0)$. Hence $\hat{x}(t)$ is rewritten

$$\hat{x}(t) = x_*(t) + e_*(t) + \tilde{E}e(t). \tag{43'}$$

The cost criterion $\hat{J}$ in (42′c) becomes, by virtue of (44), (32), and (43′),

$$\hat{J} = \sum_{t=0}^{\infty} \{(x_*(t) + e_*(t))^T\Xi(x_*(t) + e_*(t)) + (x_*(t) + e_*(t) + \tilde{E}e(t))^T K^T\Phi K(x_*(t) + e_*(t) + \tilde{E}e(t))\}. \tag{42''c}$$

J^* in (42′b) can be rewritten as

$$\begin{aligned} J^* &= x_*^T(0)Sx_*(0) \\ &= \sum_{t=0}^{\infty}\left\{x_*^T(t)Sx_*(t) - x_*^T(t+1)Sx_*(t+1)\right\} \\ &= \sum_{t=0}^{\infty}\left\{x_*^T(t)\left[S-(A-BK)^TS(A-BK)\right]x_*(t)\right\} \qquad \text{(in view of (41))} \\ &= \sum_{t=0}^{\infty}\left\{x_*^T(t)\left[\Xi + K^T\Phi K\right]x_*(t)\right\} \qquad \left(\text{in view of } (28^\dagger) \text{ in Sec. 2.2}\right). \end{aligned} \tag{42″b}$$

Subtracting J^* in (42″b) from $\hat{J}$ in (42″c) yields the value of cost increase ΔJ:

$$\begin{aligned} \Delta J = \sum_{t=0}^{\infty}\Big\{&2x_*^T(t)\Xi e_*(t) + 2\big(x_*(t)+e_*(t)\big)^TK^T\Phi K\big(e_*(t)+\tilde{E}e(t)\big) \\ &+ e_*^T(t)\left[\Xi - K^T\Phi K\right]e_*(t) + e^T(t)\tilde{E}^TK^T\Phi K\tilde{E}e(t)\Big\}. \end{aligned} \tag{45}$$

When feedback control incorporates a Luenberger observer as in (32), we have the associated behavior of state vector governed by (33′) with the property of $e(t)$ given by (34′). Subtracting (41) from (33′) with (44) taken into consideration yields

$$e_*(t) = \left[A - BK\right]e_*(t-1) - BK\tilde{E}e(t-1). \tag{46}$$

Hence

$$BK\big(e_*(t) + \tilde{E}e(t)\big) = Ae_*(t) - e_*(t+1). \tag{46′}$$

On the other hand, the Riccati equation ($28^\dagger$) and gain matrix ($27^\dagger$) in Section 2.2 give, respectively.

$$\Xi = S - \left[A - BK\right]^TSA \tag{47}$$

and

$$K^T\Phi = \left[A - BK\right]^TSB. \tag{48}$$

From (48) and (47) follows

$$\begin{aligned} \Xi - K^T\Phi K &= \Xi - \left[A - BK\right]^TSBK \\ &= S - \left[A - BK\right]^TS\left[A + BK\right]. \end{aligned} \tag{49}$$

Substituting (47) for Ξ into the first term, (48) for $K^T\Phi$ and then (46′) into the second term, and (49) for $\Xi - K^T\Phi K$ into the third term, respectively, in the brackets on the right-hand side of (45), we get

$$\Delta J = \sum_{t=0}^{\infty} \Big\{ 2x_*^T(t)\big[S - [A - BK]^T SA\big] e_*(t) + 2\big(x_*^T(t) + e_*^T(t)\big)[A - BK]^T S\big(Ae_*(t) - e_*(t+1)\big) + e_*^T(t)\big[S - [A - BK]^T S[A + BK]\big] e_*(t) + e^T(t)\tilde{E}^T K^T \Phi K \tilde{E} e(t) \Big\},$$

which reduces to

$$\Delta J = \sum_{t=0}^{\infty} \Big\{ e_*^T(t)\big[S + [A - BK]^T S[A - BK]\big] e_*(t) - 2e_*^T(t)[A - BK]^T Se_*(t+1) + e^T(t)\tilde{E}^T K^T \Phi K \tilde{E} e(t) \Big\} \qquad (45')$$

since, in view of (41) and $e_*(0) = 0$,

$$\begin{aligned} \sum_{t=0}^{\infty} & \big\{ x_*^T(t) Se_*(t) - x_*^T(t)[A - BK]^T Se_*(t+1) \big\} \\ &= \sum_{t=0}^{\infty} \big\{ x_*^T(t) Se_*(t) - x_*^T(t+1) Se_*(t+1) \big\} \\ &= x_*^T(0) Se_*(0) \\ &= 0. \end{aligned}$$

Next, the last term in the brackets on the right-hand side of (45′) is rewritten as follows with (48) taken into account:

$$\begin{aligned} e^T(t)\tilde{E}^T K^T \Phi K \tilde{E} e(t) &= e^T(t)\tilde{E}^T [A - BK]^T SBK\tilde{E} e(t) \\ &= e^T(t)\tilde{E}^T A^T SBK\tilde{E} e(t) - e^T(t)\tilde{E}^T K^T B^T SBK\tilde{E} e(t). \end{aligned}$$

The second term on the right-hand side of this equation is expressed as

$$\begin{aligned} - e^T(t)&\tilde{E}^T K^T B^T SBK\tilde{E} e(t) \\ &= \big(e_*^T(t)[A - BK]^T - e_*^T(t+1)\big) S\big(e_*(t+1) - [A - BK] e_*(t)\big) \end{aligned}$$

in view of (46). Thus ΔJ in (45′) is further reduced to

$$\begin{aligned} \Delta J &= \sum_{t=0}^{\infty} \big\{ e_*^T(t) Se_*(t) - e_*^T(t+1) Se_*(t+1) + e^T(t)\tilde{E}^T A^T SBK\tilde{E} e(t) \big\} \\ &= \sum_{t=0}^{\infty} \big\{ e^T(t)\tilde{E}^T K^T [\Phi + B^T SB] K\tilde{E} e(t) \big\}, \end{aligned} \qquad (50)$$

where $e_*(0) = 0$ and (48) have been taken into consideration. Now (50) implies that the cost $\hat{J}$ resulting from the optimal control law (32) incorporating a Luenberger observer will, in general, be greater than the standard minimal cost J^* in (42′b), provided estimation error $e(t)$ is nonzero for some t.

Finally, since $e(t) = Fe(t-1)$ in (34′), we have

$$\Delta J = e^T(0)\sum_{t=0}^{\infty}(K\tilde{E}F^t)^T[\Phi + B^TSB]K\tilde{E}F^t e(0), \tag{50′}$$

implying that only if the initial estimation error $e(0)$ were zero, i.e., $\hat{w}(0) = w(0)$, the related cost function would assume the minimum value J^*. However, this is simply the case where all state variables are really accessible, so as long as we use an observer, the cost $\hat{J}$ associated with the optimal control (32) will be greater than J^*. (For a parallel argument, refer to Gourishankar and Kudva(1977).)

3.4. Recursive Minimum-Cost Observer

In the preceding sections we know that many Luenberger observers exist and that they definitely incur cost increase. We shall try to obtain an observer which incurs the least cost increment; our approach is a recursive type of estimation which has some relationship to the Kalman's estimation developed in Chapter 4 for stochastic systems.

We consider the system consisting of (1) and (3):

$$x(t) = Ax(t-1) + Bv(t) \tag{1}$$

$$y(t) = Cx(t), \tag{3}$$

where A, B, and C are real constant matrices of dimensions $n \times n$, $n \times m$, and $r \times n$, respectively, with $rk(C) = r(< n)$. Tse and Athans (1970) give the following definition.

Definition 1. Let C be an $r \times n$ real constant matrix with rank r and let T denote any $s \times n$ real matrix with $s = n - r$. The set

$$\Upsilon(C) = \{T : N(T) \cap N(C) = \emptyset\} \qquad (\emptyset \text{ denotes an empty set.}) \tag{51}$$

is called the set of complementary matrices of order $n - r$ for C, where $N(C)$ and $N(T)$ denote the nullspaces of C and T, respectively.

By virtue of this definition, the only vector b satisfying both $Cb = 0$ and $Tb = 0$ is a null vector; i.e. (denoting transposition by superscript T), $n \times n$ matrix $[C^T, T^T]^T$ is nonsingular for each $T \in \Upsilon(C)$ in view of Theorem 27 in Murata (1977, Section 2.3). Thus we have an $n \times r$ matrix $Q(t)$ and an $n \times s$ matrix $P(t)$ such that

$$Q(t)C + P(t)T(t) = I \qquad \text{for} \quad T(t) \in \Upsilon(C), \tag{52}$$

where $[Q(t), P(t)]$ is the inverse of $[C^T, T^T(t)]^T$.

Theorem 4 (Tse and Athans, 1970). *Let* $\{T\} \equiv \{T(0), T(1), T(2), \ldots, T(\infty)\}$ *be a sequence of matrices* $T(t) \in \Upsilon(C)$. *Then, for system* (1), (3), *an order-s observer*

$$z(t) = F(t)z(t-1) + H(t)y(t-1) + T(t)Bv(t) \tag{53}$$

exists for $t = 1, 2, \ldots$ *such that for some appropriate choice of* $z(0)$

$$z(t) = T(t)x(t), \qquad t = 1, 2, \ldots . \tag{54}$$

Proof. Pick for $t = 1, 2, 3, \ldots$

$$F(t) = T(t)AP(t-1) \tag{55}$$

$$H(t) = T(t)AQ(t-1), \tag{56}$$

where $Q(t)$ and $P(t)$ satisfy (52). Substituting (55), (56), and (3) into (53), and subtracting $T(t)x(t)$, with (1) taken into account, from the resulting equation we get

$$\begin{aligned} z(t) - T(t)x(t) = {} & T(t)AP(t-1)z(t-1) + T(t)AQ(t-1)Cx(t-1) \\ & - T(t)Ax(t-1), \end{aligned}$$

and then (52) is taken into account:

$$z(t) - T(t)x(t) = T(t)AP(t-1)\{z(t-1) - T(t-1)x(t-1)\}. \tag{57}$$

Thus if we choose $z(0) = T(0)x(0)$, then we obtain (54). □

Remark. From the above proof, we know that the observer $z(t)$ in (53) can be described in terms of the sequence $\{T\}$ as

$$z(t) = T(t)A\{P(t-1)z(t-1) + Q(t-1)y(t-1)\} + T(t)Bv(t). \tag{58}$$

Also, by defining

$$e(t) \equiv z(t) - T(t)x(t), \tag{59}$$

we have from (57)

$$e(t) = F(t)e(t-1). \tag{60}$$

Combining (53) with (59) and taking (60) into account yield

$$\begin{aligned} & F(t)T(t-1)x(t-1) + H(t)Cx(t-1) + T(t)Bv(t) \\ & \quad = T(t)\{Ax(t-1) + Bv(t)\} \end{aligned}$$

or, equivalently,

$$T(t)A - F(t)T(t-1) = H(t)C \tag{61}$$

since $x(t)$ is nontrivial. Equation (61) is a dynamic version of the fundamental observer equation (19). We can easily show, in a manner similar to Lemma 2, that if $T(t)$ satisfies (61) then (60) follows from (53). Therefore, we have the following.

Theorem 5. *Let* $\{T\} \equiv \{T(0), T(1), T(2), \ldots, T(\infty)\}$ *be a sequence of matrices* $T(t) \in \Upsilon(C)$. *Then equations* (60) *and* (61) *hold and are equivalent to each other.*

The proof follows immediately from Theorem 4 and the above remark. Note that (52), (55), and (56) imply (61) since

$$\begin{aligned} &F(t)T(t-1) + H(t)C \\ &\quad = T(t)A\{P(t-1)T(t-1) + Q(t-1)C\} = T(t)A, \end{aligned}$$

and that, conversely, (52) and (61) imply (55) and (56). Equation (52) follows from $T(t) \in \Upsilon(C)$. Thus Theorem 5 may alternatively be stated as follows.

Theorem 5′. *Let* $\{T\} \equiv \{T(0), T(1), T(2), \ldots, T(\infty)\}$ *be a sequence of matrices* $T(t) \in \Upsilon(C)$. *Then* (55), (56), *and* (60) *hold, and* (55), (56) *are equivalent to* (60).

In the following, we derive an explicit form of $T(t)$, which should belong to $\Upsilon(C)$ since we utilize (55), (56), and (60).

Now we want to minimize the sum of squares of estimation error $e(t)$ (cf. (35′)), by which we may minimize the accompanying cost increase (cf. (50)). The sum of the squares of the estimation error is defined as $e^T(t)e(t)$, which is equal to the trace of $e(t)e^T(t)$, the covariance matrix of $e(t)$. By virtue of (60), the covariance matrix is written in a recursive form:

$$e(t)e^T(t) = T(t)\Omega(t-1)T^T(t), \tag{62}$$

where

$$\Omega(t) \equiv AP(t)e(t)e^T(t)P^T(t)A^T. \tag{63}$$

Conformably to the partition of A in (4), we partition $\Omega(t)$ as

$$\Omega(t) = \begin{pmatrix} \Omega_{11}(t) & \Omega_{12}(t) \\ \Omega_{21}(t) & \Omega_{22}(t) \end{pmatrix} \tag{64}$$

where

$$\begin{aligned} &\Omega_{ii}(t) \equiv [A_{i1}, A_{i2}]P(t)e(t)e^T(t)P^T(t)[A_{i1}, A_{i2}]^T \qquad \text{for} \quad i = 1,2; \\ &\Omega_{12}(t) = \Omega_{21}^T(t) \equiv [A_{11}, A_{12}]P(t)e(t)e^T(t)P^T(t)[A_{21}, A_{22}]^T. \end{aligned} \tag{65}$$

(For a parallel argument on a stochastic system, refer to Leondes and Novak (1972).)

Assuming that the first r vector of x is observed, we specify C as

$$C = [I_r, 0]. \tag{66}$$

Correspondingly, $T(t)$ is partitioned as (cf. (21))

$$T(t) = [-G(t), I_s], \tag{67}$$

where $G(t)$ is an $s \times r$ matrix function of t. Then

$$\begin{pmatrix} C \\ T(t) \end{pmatrix}^{-1} = \begin{pmatrix} I_r & 0 \\ -G(t) & I_s \end{pmatrix}^{-1} = \begin{pmatrix} I_r & 0 \\ G(t) & I_s \end{pmatrix}, \tag{68}$$

i.e.,

$$Q(t) = \begin{pmatrix} I_r \\ G(t) \end{pmatrix} \quad \text{and} \quad P(t) = \begin{pmatrix} 0 \\ I_s \end{pmatrix} \equiv \tilde{E}. \tag{69}$$

Hence $\Omega_{ij}(t)$ in (65) reduce to

$$\begin{aligned} \Omega_{ii}(t) &= A_{i2}e(t)e^T(t)A_{i2}^T \qquad \text{for} \quad i = 1,2; \\ \Omega_{12}(t) &= \Omega_{21}^T(t) = A_{12}e(t)e^T(t)A_{22}^T, \end{aligned} \tag{65'}$$

and the covariance matrix of $e(t)$ is expressed as

$$\begin{aligned} e(t)e^T(t) = G(t)\Omega_{11}(t-1)G^T(t) &- G(t)\Omega_{12}(t-1) \\ &- \Omega_{21}(t-1)G^T(t) + \Omega_{22}(t-1). \end{aligned} \tag{62'}$$

Thus the trace of the covariance matrix is quadratic in $G(t)$. The extremum condition for the trace is given by setting its gradient with respect to $G(t)$ equal to zero:

$$\begin{aligned} 0 &= \frac{\partial \operatorname{tr}\big(e(t)e^T(t)\big)}{\partial G(t)} \\ &= 2G(t)\Omega_{11}(t-1) - \Omega_{12}^T(t-1) - \Omega_{21}(t-1) \\ &= 2G(t)\Omega_{11}(t-1) - 2\Omega_{21}(t-1), \end{aligned}$$

from which follows (for singular Ω_{11}, its generalized inverse replaces Ω_{11}^{-1})

$$G(t) = \Omega_{21}(t-1)\Omega_{11}(t-1)^{-1}. \tag{70}$$

Note that this extremum condition is also sufficient for $\operatorname{tr}(e(t)e^T(t))$ to achieve a minimum since Ω_{11} is positive semidefinite. Substituting (70) into (62′) yields

$$e(t)e^T(t) = \Omega_{22}(t-1) - \Omega_{21}(t-1)\Omega_{11}(t-1)^{-1}\Omega_{12}(t-1). \tag{71}$$

Formula (70) indicates a recursive procedure for obtaining an optimal observer. Substitution of (67) and (69) into (55) and (56) yields

$$F(t) = A_{22} - G(t)A_{12}, \tag{55'}$$

$$H(t) = A_{21} - G(t)A_{11} + (A_{22} - G(t)A_{12})G(t-1). \tag{56'}$$

Thus we reduce the explicit form of observer in (58) to

$$\begin{aligned} z(t) = (A_{22} &- G(t)A_{12})z(t-1) + (B_{22} - G(t)B_{11})v(t) \\ &+ \big[A_{21} - G(t)A_{11} + (A_{22} - G(t)A_{12})G(t-1)\big]y(t-1), \end{aligned} \tag{58'}$$

which may be termed a recursive observer, as compared with $z(t)$ in (25).

Initiation of the observer can be done as follows. Let

$$z(0)=\bar{w}(0), \tag{72}$$

where $\bar{w}(0)$ is some expected value of unobserved s vector $w(0)$ of $x(0)$ in period 0. Equation (72) implies

$$G(0)=0 \quad \text{and hence} \quad T(0)=\left[0, I_s\right]. \tag{73}$$

Since $e(0)=z(0)-T(0)x(0)=\bar{w}(0)-w(0)$,

$$e(0)e^T(0)=(\bar{w}(0)-w(0))(\bar{w}^T(0)-w^T(0)) \equiv W_0. \tag{74}$$

Hence

$$\begin{pmatrix} \Omega_{11}(0) & \Omega_{12}(0) \\ \Omega_{21}(0) & \Omega_{22}(0) \end{pmatrix}=\begin{pmatrix} A_{12}W_0A_{12}^T & A_{12}W_0A_{22}^T \\ A_{22}W_0A_{12}^T & A_{22}W_0A_{22}^T \end{pmatrix}. \tag{75}$$

Then by (70)

$$G(1)=\Omega_{21}(0)\Omega_{11}(0)^{-1}=A_{22}W_0A_{12}^T\left(A_{12}W_0A_{12}^T\right)^{-1} \tag{70'}$$

and by (71)

$$\begin{aligned} W_1 \equiv e(1)e^T(1) &= \Omega_{22}(0)-\Omega_{21}(0)\Omega_{11}(0)^{-1}\Omega_{12}(0) \\ &= A_{22}\left[W_0-W_0A_{12}^T\left(A_{12}W_0A_{12}^T\right)^{-1}A_{12}W_0\right]A_{22}^T. \end{aligned} \tag{71'}$$

For $t=2$, we substitute (71′) into (65) and the result into (70) and (71) to obtain $G(2)$ and $e(2)e^T(2)$, and so forth. Progressing, one by one, in a similar way we can calculate all $G(t)$ and $e(t)e^T(t)$, $t=1,2,\ldots$; that is, for $t=1,2,3,\ldots$ (if $A_{12}W_{t-1}A_{12}^T$ is singular, its generalized inverse will be as below),

$$G(t)=A_{22}W_{t-1}A_{12}^T\left(A_{12}W_{t-1}A_{12}^T\right)^{-1}, \tag{70†}$$

$$\begin{aligned} W_t &\equiv e(t)e^T(t) \\ &= A_{22}\left[W_{t-1}-W_{t-1}A_{12}^T\left(A_{12}W_{t-1}A_{12}^T\right)^{-1}A_{12}W_{t-1}\right]A_{22}^T. \end{aligned} \tag{71†}$$

Finally, it is easy to show that $T(0), T(1), T(2), \ldots$ thus obtained belong to $\Upsilon(C)$ since for $t=0$

$$\begin{pmatrix} C \\ T(0) \end{pmatrix} b=\begin{pmatrix} I_r & 0 \\ 0 & I_s \end{pmatrix} b=0 \Rightarrow b=0,$$

and for $t=1,2,3,\ldots$

$$\begin{pmatrix} C \\ T(t) \end{pmatrix}\begin{pmatrix} b_r \\ b_s \end{pmatrix} \equiv \begin{pmatrix} I_r & 0 \\ -G(t) & I_s \end{pmatrix}\begin{pmatrix} b_r \\ b_s \end{pmatrix}=0 \Rightarrow \begin{cases} b_r=0 \\ b_s=G(t)b_r=0, \end{cases}$$

where b_r and b_s stand for the first r vector and the last s vector of n vector b, respectively.

Note also that in our recursive procedure

$$F(0) = A_{22},$$

$$F(t) = A_{22} - G(t)A_{12} = A_{22} - \Omega_{21}(t-1)\Omega_{11}(t-1)^{-1}A_{12} \tag{76}$$

$$= A_{22} - A_{22}W_{t-1}A_{12}^T\left(A_{12}W_{t-1}A_{12}^T\right)^{-1}A_{12} \tag{77}$$

for $t = 1, 2, \ldots$. If $e(t)$ were zero for $t = 0, 1, 2, \ldots$, therefore, no common eigenvalue would exist for A_{22} and for $F(t)$.

As for the stability property of the system incorporating our dynamic observer, we may consider the composite system

$$\begin{pmatrix} x(t) \\ e(t) \end{pmatrix} = \begin{pmatrix} A - BK & -BK\tilde{E} \\ 0 & F(t) \end{pmatrix} \begin{pmatrix} x(t-1) \\ e(t-1) \end{pmatrix}, \tag{78}$$

which can be derived in the same way as system (36). The stability of system (78) depends on that of $F(t)$ determined recursively by (77). For the related stability analysis, refer to Tse (1973).

On the assumption that system (78) is stable, the observer $\hat{w}(t)$ for unobserved state vector $w(t)$ is defined as

$$\hat{w}(t) \equiv z(t) + G(t)y(t) \tag{79}$$

since then $w(t)$ is expressed as

$$w(t) = \hat{w}(t) + F(t)\{w(t-1) - \hat{w}(t-1)\} \tag{80}$$

in view of (57) and (67). The observer $\hat{x}(t)$ for the whole state vector $x(t)$ is written

$$\hat{x}(t) = Q(t)Cx(t) + \tilde{E}z(t) \tag{81}$$

since

$$\begin{pmatrix} y(t) \\ \hat{w}(t) \end{pmatrix} = \begin{pmatrix} I_r & 0 \\ G(t) & I_s \end{pmatrix} \begin{pmatrix} y(t) \\ z(t) \end{pmatrix}.$$

An apparent difference between the present minimum-cost observer and the Luenberger observer discussed in Section 3.3 is that, while the latter involves a static matrix G obtained through rather cumbersome steps, the corresponding matrix $G(t)$ in the minimum-cost observer is easily calculated in a recursive manner.

Application 1. We consider a macroeconomic model in which the consumption is linearly dependent on permanent income and the GNP. Our model is as follows:

$$q(t) = c(t) + i(t) + v(t) \tag{82a}$$

$$c(t) = \alpha_1 z(t) + \alpha_2 q(t-1) \tag{82b}$$

$$i(t) = \alpha_3 q(t) - \alpha_4 k(t-1) \tag{82c}$$

$$k(t) = k(t-1) + i(t) \tag{82d}$$

$$z(t) = \alpha_5 z(t-1) + \alpha_6 i(t), \tag{82e}$$

where we denote z = permanent income, q = GNP, c = consumption, i = investment, v = government expenditures, k = capital stock, and α's are positive constants. Equations (82a) and (82d) are definitional ones, (82b) shows the consumption behavior, (82c) represents an acceleration principle of investment, and (82e) determines the movement of permanent income. (Constant terms are omitted for the sake of brevity.) We assume that $z(t)$ cannot be directly observed and that $v(t)$ is a control variable (instrument). Substitutions of (82b) and (82c) into (82a) and of (82c) into (82d) and (82e) yield, respectively,

$$(1-\alpha_3)q(t)-\alpha_1 z(t)=\alpha_2 q(t-1)-\alpha_4 k(t-1)+v(t),$$
$$k(t)-\alpha_3 q(t)=(1-\alpha_4)k(t-1),$$
$$z(t)-\alpha_3\alpha_6 q(t)=\alpha_5 z(t-1)-\alpha_4\alpha_6 k(t-1).$$

These condensed equations are expressed in matrix form

$$\begin{bmatrix} 1-\alpha_3 & 0 & -\alpha_1 \\ -\alpha_3 & 1 & 0 \\ -\alpha_3\alpha_6 & 0 & 1 \end{bmatrix}\begin{bmatrix} q(t) \\ k(t) \\ z(t) \end{bmatrix}=\begin{bmatrix} \alpha_2 & -\alpha_4 & 0 \\ 0 & 1-\alpha_4 & 0 \\ 0 & -\alpha_4\alpha_6 & \alpha_5 \end{bmatrix}\begin{bmatrix} q(t-1) \\ k(t-1) \\ z(t-1) \end{bmatrix}+\begin{bmatrix} 1 \\ 0 \\ 0 \end{bmatrix}v(t). \tag{83}$$

Premultiplying (83) by the inverse of the coefficient matrix on its left-hand side, we obtain the state-space form of our model corresponding to (1)

$$\begin{bmatrix} q(t) \\ k(t) \\ z(t) \end{bmatrix}=\begin{bmatrix} a_{11} & a_{12} & a_{13} \\ a_{21} & a_{22} & a_{23} \\ a_{31} & a_{32} & a_{33} \end{bmatrix}\begin{bmatrix} q(t-1) \\ k(t-1) \\ z(t-1) \end{bmatrix}+\begin{bmatrix} \rho \\ \rho\alpha_3 \\ \rho\alpha_3\alpha_6 \end{bmatrix}v(t), \tag{83'}$$

where $\rho\equiv(1-\alpha_3-\alpha_1\alpha_3\alpha_6)^{-1}$ and

$$\begin{bmatrix} a_{11} & a_{12} & a_{13} \\ a_{21} & a_{22} & a_{23} \\ a_{31} & a_{32} & a_{33} \end{bmatrix}\equiv\begin{bmatrix} \rho & 0 & \rho\alpha_1 \\ \rho\alpha_3 & 1 & \rho\alpha_1\alpha_3 \\ \rho\alpha_3\alpha_6 & 0 & \rho(1-\alpha_3) \end{bmatrix}\begin{bmatrix} \alpha_2 & -\alpha_4 & 0 \\ 0 & 1-\alpha_4 & 0 \\ 0 & -\alpha_4\alpha_6 & \alpha_5 \end{bmatrix}. \tag{84}$$

Since $z(t)$ is the only unobserved variable, coefficient matrix C in output equation (3) becomes in this case

$$C\equiv\begin{pmatrix} 1 & 0 & 0 \\ 0 & 1 & 0 \end{pmatrix}.$$

Then, in view of (66), we can apply ($70^\dagger$) and (58′) to the present model.

Since W_{t-1} in ($70^\dagger$) is a scalar in this example, and since $A_{12}A_{12}^T$ is singular, $G(t)$ reduces to

$$G(t)=A_{22}A_{12}^T\left(A_{12}A_{12}^T\right)^+ \tag{85}$$

where $(A_{12}A_{12}^T)^+$ is the generalized inverse obtained by applying Theorem

22 in Section 6.3 of Murata (1977, p. 215), which is calculated as

$$(A_{12}A_{12}^T)^+ = (A_{12}^T)^+ A_{12}^+ = \frac{1}{(a_{13}^2 + a_{23}^2)^2}\begin{pmatrix} a_{13}^2 & a_{13}a_{23} \\ a_{13}a_{23} & a_{23}^2 \end{pmatrix}$$

in view of the partitions

$$\begin{pmatrix} A_{11} & \vdots & A_{12} \\ \cdots & & \cdots \\ A_{21} & \vdots & A_{22} \end{pmatrix} = \begin{bmatrix} a_{11} & a_{12} & \vdots & a_{13} \\ a_{21} & a_{22} & \vdots & a_{23} \\ \cdots & \cdots & & \cdots \\ a_{31} & a_{32} & \vdots & a_{33} \end{bmatrix}, \qquad \begin{pmatrix} B_{11} \\ \cdots \\ B_{22} \end{pmatrix} = \begin{bmatrix} \rho \\ \rho\alpha_3 \\ \cdots \\ \rho\alpha_3\alpha_6 \end{bmatrix}.$$

Thus $G(t)$ in (85) becomes a two-dimensional vector of constant components (g_1, g_2) such as

$$G(t) = \left(\frac{a_{13}a_{33}}{a_{13}^2 + a_{23}^2}, \frac{a_{23}a_{33}}{a_{13}^2 + a_{23}^2} \right) \equiv (g_1, g_2),$$

and hence we have $A_{22} - G(t)A_{12} = 0$. Therefore, our observer $z(t)$ (as in (58′)) reduces to

$$\begin{aligned} z(t) &= (B_{22} - G(t)B_{11})v(t) + (A_{21} - G(t)A_{11})\begin{pmatrix} q(t-1) \\ k(t-1) \end{pmatrix} \\ &= \rho(\alpha_3\alpha_6 - g_1 - \alpha_3 g_2)v(t) + (a_{31} - a_{11}g_1 - a_{21}g_2)q(t-1) \\ &\quad + (a_{32} - a_{12}g_1 - a_{22}g_2)k(t-1). \end{aligned} \tag{86}$$

Utilizing the definitions of (84) and ρ, we can express $z(t)$ in (86) as

$$z(t) = -\frac{1}{\alpha_1}v(t) - \frac{\alpha_2}{\alpha_1}q(t-1) + \left\{ \frac{1}{\alpha_1} - \frac{(1-\alpha_4)(1+\alpha_3)}{\alpha_1(1+\alpha_3^2)} \right\} k(t-1). \tag{86′}$$

3.5. Separation of Observer and Controller

Our objective in this chapter is to obtain observers as proxies for inaccessible state variables in order to apply them to the optimal feedback control rules developed separately, as in Section 2.2. From this standpoint, we have tried to minimize the cost rise due to incorporating observers into a controller, assuming the controller design as given. However, since the controller should be designed to minimize the associated cost, it seems more appropriate to compute the control and observer parameters simultaneously in such a way that the overall cost is minimized. We shall consider this design procedure in the present section, following Willems (1980) who confirms our view of the separation of observer and controller design for a linear discrete-time optimal control problem.

We want to design the control input $v(t)$ to minimize the cost

$$J = \sum_{t=1}^{\infty} \{x^T(t-1)\Xi x(t-1) + v^T(t)\Phi v(t)\} \tag{87}$$

subject to the linear system

$$x(t) = Ax(t-1) + Bv(t), \tag{1}$$

where x is an n vector of state variables, v is an m vector of control variables, Ξ and Φ are positive definite matrices, and A, B are constant matrices of appropriate dimensions. We assume that only a limited number r $(< n)$ of state variables are measured, and we represent them by the output vector y as

$$y(t) = Cx(t) \qquad \text{with} \quad C = [I_r, 0]. \tag{3}$$

If all the state variables were measurable, the optimal input would be (cf. Thoerem 15 in Section 2.2)

$$v(t) = -Kx(t-1) \tag{88}$$

with

$$K \equiv [B^TSB + \Phi]^{-1}B^TSA, \tag{89}$$

where K is the unique positive definite solution of the algebraic Riccati equation:

$$S = A^T\Big[S - SB[B^TSB + \Phi]^{-1}B^TS\Big]A + \Xi. \tag{90}$$

Substituting (89) into (90) yields

$$S = A^TS[A - BK] + \Xi. \tag{90$'$}$$

Also, it follows from (89) that

$$B^TSA = [B^TSB + \Phi]K. \tag{89$'$}$$

Then Ξ in (87) is eliminated as follows, with (90′), (1), and (89′) taken into consideration:

$$\begin{aligned} J &= \sum_{t=1}^{\infty}\{x^T(t)Sx(t) + x^T(t-1)A^TS[BK - A]x(t-1) + v^T(t)\Phi v(t)\} \\ &\quad + x^T(0)Sx(0) \\ &= \sum_{t=1}^{\infty}\{x^T(t-1)A^TSB(Kx(t-1) + v(t)) + v^T(t)B^TSAx(t-1) \\ &\qquad + v^T(t)[B^TSB + \Phi]v(t)\} + x^T(0)Sx(0) \\ &= \sum_{t=1}^{\infty}\{(v(t) + Kx(t-1))^T[B^TSB + \Phi](v(t) + Kx(t-1))\} \\ &\quad + x^T(0)Sx(0) \end{aligned} \tag{87$'$}$$

where K is the matrix defined by (89).

When outputs $y(\tau)$ ($\tau = 0, 1, \ldots, t$) only are available to us, we must reconstruct the state $x(t)$. In this case the estimate is denoted by $\hat{x}(t)$. The design of optimal state reconstruction consists of two parts: first, to design the optimal control

$$v(t+1) = -\widehat{K\hat{x}}(t) \tag{91}$$

where $\widehat{K\hat{x}}(t)$ is an optimal estimate of $Kx(t)$, using the available information as of time t; secondly, in view of the cost criterion in (87′), to design the estimate $\hat{x}(t)$ such that the estimation error measure

$$\rho \equiv \sum_{t=0}^{\infty} (\widehat{K\hat{x}}(t) - Kx(t))^T [B^T S B + \Phi](\widehat{K\hat{x}}(t) - Kx(t)) \tag{92}$$

is to be minimized. Of these two, the control design problem is obviously independent of the estimation, since the computation of the gain matrix K depends only on the parameters defining the system (1) and the cost function (87). Hence the optimal control can be expressed as

$$v(t+1) = -K\hat{x}(t) \tag{91′}$$

instead of (91). However, the estimation of $\hat{x}(t)$ may depend on the control, since the estimation error criterion ρ in (92) involves matrices K and Φ. We therefore analyze the design of optimal observers with respect to the estimation error measure

$$\rho \equiv \sum_{t=0}^{\infty} (\hat{x}(t) - x(t))^T M (\hat{x}(t) - x(t)) \tag{92′}$$

where $M \equiv K^T[B^T S B + \Phi]K$, and we show that the optimal observer design is independent of the matrix M and thus independent of the controller design.

We start with the construction of a full-order observer from the minimal-order Luenberger observer described by (25) and (30). $\hat{x}(t)$ in (31′) of Section 3.2 can be expressed as

$$\hat{x}(t) = Qy(t) + r(t) \tag{93}$$

where $r(t) \equiv \tilde{E}z(t)$, and Q, $\hat{E}$ are those in (31′). Premultiplying both sides of the observer equation (25) by $\tilde{E}$, and taking the partitions in (4) into account, we get

$$r(t) = [A - QCA]r(t-1) + [A - QCA]\,Qy(t-1) + [B - QCB]v(t) \tag{94}$$

where C is that in (3). We shall try to design matrix Q in the full-order observer (93), (94), such that the estimation performance measure ρ is minimized.

Defining the estimation error $\hat{\epsilon}$ of the state x as

$$\hat{\epsilon}(t) \equiv x(t) - \hat{x}(t), \tag{95}$$

we can derive the following relation (96), in view of (1), (3), (93), and (94),

$$\hat{\epsilon}(t) = [A - QCA]\hat{\epsilon}(t-1) \qquad \text{for} \quad t = 1, 2, \ldots . \tag{96}$$

Given the initial values of state $x(0)$ and *a priori* estimate $\hat{x}(0)$, the initial estimation error $\hat{\epsilon}(0)$ is determined as

$$\hat{\epsilon}(0) = x(0) - \hat{x}(0) = x(0) - Qy(0) - r(0), \tag{97}$$

and the initial-stage observer $r(0)$ is, in turn, assumed to be equal to

$$r(0) = \hat{x}(0) - Q\hat{y}(0), \tag{98}$$

where $\hat{y}(0) = C\hat{x}(0)$. Then, substituting (98) into (97) yields

$$\hat{\epsilon}(0) = [I - QC](x(0) - \hat{x}(0)). \tag{97$'$}$$

By denoting

$$\Theta \equiv (x(0) - \hat{x}(0))(x(0) - \hat{x}(0))^T, \tag{99}$$

we have

$$\hat{\epsilon}(0)\hat{\epsilon}^T(0) = [I - QC]\Theta[I - QC]^T \equiv \hat{\Theta}. \tag{100}$$

On the other hand, it follows from (96) that

$$\hat{\epsilon}(t)\hat{\epsilon}^T(t) = [I - QC]A\hat{\epsilon}(t-1)\hat{\epsilon}^T(t-1)A^T[I - QC]^T$$
$$\text{for} \quad t = 1, 2, \ldots . \tag{101}$$

The estimation performance criterion ρ is now expressed as

$$\begin{aligned} \rho &= \sum_{t=0}^{\infty} \left(\hat{\epsilon}^T(t) M \hat{\epsilon}(t)\right) \\ &= \operatorname{tr}\left(M \sum_{t=0}^{\infty} \hat{\epsilon}(t)\hat{\epsilon}^T(t)\right) \\ &= \operatorname{tr}(M\Pi), \end{aligned} \tag{102}$$

where

$$\Pi \equiv \sum_{t=0}^{\infty} \left([I - QC]A\right)^t \hat{\Theta} \left(A^T[I - QC]^T\right)^t \tag{103}$$

in view of (100) and (101). Clearly, Π is the symmetric positive semidefinite solution of the following algebraic equation:

$$\Pi = \hat{A}\Pi\hat{A}^T + \hat{\Theta}, \tag{104}$$

where

$$\hat{A} \equiv [I - QC]A. \tag{105}$$

Our problem is now reduced to choosing Q and Π so as to minimize ρ under the constraint (104). This constrained problem can be reformulated as the unconstrained problem of choosing Q, Π, and Λ so as to minimize the Lagrangian

$$L = \operatorname{tr}(M\Pi) + \operatorname{tr}(\Lambda[\hat{A}\Pi\hat{A}^T + \hat{\Theta} - \Pi]), \tag{106}$$

where Λ is an $n \times n$ matrix of Lagrange multipliers. (Cf. Application 3 Murata (1977, p. 262).) Note that as matrix $[\hat{A}\Pi\hat{A}^T + \hat{\Theta} - \Pi]$ is symmetric, so is Λ. Necessary conditions for optimality are

$$\frac{\partial L}{\partial Q} = 0, \qquad \frac{\partial L}{\partial \Pi} = 0, \tag{107}$$

and (104). More specifically, conditions (107) are expressed as follows (see the Appendix):

$$\begin{aligned}\frac{\partial L}{\partial Q} &= \frac{\partial \operatorname{tr}(\Lambda[I - QC]A\Pi A^T[I - QC]^T + \Lambda[I - QC]\Theta[I - QC]^T)}{\partial Q} \\ &= 2\Lambda[QC - I][A\Pi A^T C^T + \Theta C^T] = 0, \end{aligned} \tag{108}$$

and

$$\begin{aligned}\frac{\partial L}{\partial \Pi} &= \frac{\partial \operatorname{tr}([M - \Lambda]\Pi + \Lambda\hat{A}\Pi\hat{A}^T)}{\partial \Pi} \\ &= M - \Lambda + \hat{A}^T\Lambda\hat{A} = 0. \end{aligned} \tag{109}$$

While (109) indicates some interdependence among M, Λ, and Q, equation (108) implies that the optimal observer matrix Q is determined, as (110), independently of Λ and hence of M, provided Λ is nonsingular.

$$Q = A\Pi A^T C^T[CA\Pi A^T C^T]^{-1} \tag{110}$$

since $\Theta C^T = 0$ in view of $C = [I_r, 0]$ and of

$$\Theta = \begin{pmatrix} 0 & 0 \\ 0 & e(0)e^T(0) \end{pmatrix}, \tag{99$'$}$$

where $e(0)$ $(\equiv \hat{w}(0) - \mathrm{w}(0))$ is the estimation error of a minimal-order observer $\hat{w}$. Conformably with partition A in (4), we partition Π as

$$\Pi = \begin{pmatrix} \Pi_{11} & \Pi_{12} \\ \Pi_{21} & \Pi_{22} \end{pmatrix} \quad \text{with} \quad \Pi_{21}^T = \Pi_{12}, \tag{111}$$

and calculate the right-hand side of (110), entailing

$$Q = \begin{pmatrix} I_r \\ \hat{G} \end{pmatrix} \tag{110$'$}$$

where

$$\hat{G} = (A_{21}, A_{22})\Pi(A_{11}, A_{12})^T[(A_{11}, A_{12})\Pi(A_{11}, A_{12})^T]^{-1}. \tag{112}$$

Substituting (99′) for Θ and (110′) for Q into (100) shows that $\hat{\Theta}$ is equal to Θ in (99′). Similar substitutions for $\hat{\Theta}$ and Q into (103) yield $\Pi_{11} = 0$,

$\Pi_{12} = \Pi_{21}^{\mathrm{T}} = 0$ and

$$\Pi_{22} = \sum_{t=0}^{\infty} (A_{22} - \hat{G}A_{12})^t e(0)e^T(0)(A_{22}^T - A_{12}^T\hat{G}^T)^t. \tag{113}$$

Hence $\hat{G}$ in (112) reduces to

$$\hat{G} = A_{22}\Pi_{22}A_{12}^T\left[A_{12}\Pi_{22}A_{12}^T\right]^{-1}, \tag{112$'$}$$

and accordingly our full-order observer described by (93) and (94) reduces to a minimal-order Luenberger observer described by (93′) and (94′) below:

$$\hat{w}(t) = \hat{G}y(t) + z(t), \tag{93$'$}$$

$$\begin{aligned} z(t) = &\left[A_{22} - \hat{G}A_{12}\right]z(t-1) + \left[A_{21} - \hat{G}A_{11} + (A_{22} - \hat{G}A_{12})\hat{G}\right]y(t-1) \\ &+ \left[B_{22} - \hat{G}B_{11}\right]v(t). \end{aligned} \tag{94$'$}$$

The features of this result are enumerated here.

1) The optimal reconstructor of inaccessible state variables is a minimal-order Luenberger observer.
2) The optimal observer matrix $\hat{G}$ is independent of the estimation error weighing matrix M and thus independent of the optimal controller design for our linear discrete-time optimal control system.

For an analysis similar to the above separation argument, and for the design of optimal controllers based on the separation principle, refer to O'Reilly and Newmann (1976). Furthermore, by combining the minimal-order observer (93′), (94′) with a feedback controller, we recognize that the corresponding linear dynamical controller takes the form of

$$v(t) = Rz(t-1) + Uy(t-1), \tag{114}$$

$$z(t) = Dz(t-1) + Vy(t-1), \qquad z(0) = z_0. \tag{115}$$

Given this controller form, O'Reilly (1978) proposes to determine an optimal low-order controller with the dimension of $z(t)$ set arbitrarily at s in the interval of $0 \leqslant s \leqslant n - r$.

References

Aoki, M., and Huddle, J. R., "Estimation of the state vector of a linear stochastic system with a constrained estimator," *IEEE Trans. Automatic Control*, Vol. AC-12, pp. 432–433, 1967.

Arbel, A., and Tse, E., "Observer design for large-scale linear systems," *IEEE Trans. Automatic Control*, vol. AC-24, pp. 469–476, 1979.

Gopinath, B., "On the control of linear multiple input–output systems," *Bell System Technical J.*, vol. 50, pp. 1063–1081, 1971.

Gourishankar, V., and Kudva, P., "Optimal observers for the state regulation of linear discrete-time plants," *International J. Control*, vol. 26, pp. 359–368, 1977.

Leondes, C. T., and Novak, L. M., "Optimal minimal-order observers for discrete-time systems—A unified theory," *Automatica*, vol. 8, pp. 379–387, 1972.

Luenberger, D. G., "Observing the state of a linear system," *IEEE Trans. Military Electronics*, vol. 8, pp. 74–80, 1964.

——, "Observer for multivariable systems," *IEEE Trans. Automatic Control*, vol. AC-11, pp. 190–197, 1966.

——, "An introduction to observers," *IEEE Trans. Automatic Control*, vol. AC-16, pp. 596–602, 1971.

Murata, Y., *Mathematics for Stability and Optimization of Economic Systems.* New York: Academic Press, 1977.

O'Reilly, J., "Optimal low-order feedback controllers for discrete-time linear systems with inaccessible state," *International J. Systems Science*, vol. 9, pp. 301–309, 1978.

O'Reilly, J., and Newmann, M. M., "On the design of discrete-time optimal dynamical controllers using a minimal-order observer," *International J. Control*, vol. 23, pp. 257–275, 1976.

Tse, E., "Observer-estimators for discrete-time systems," *IEEE Trans. Automatic Control*, vol. AC-18, pp. 10–16, 1973.

Tse, E., and Athans, M., "Optimal minimal-order observer-estimators for discrete linear time-varying systems," *IEEE Trans. Automatic Control*, vol. AC-15, pp. 416–426, 1970.

Willems, J. L., "Optimal state reconstruction algorithms for linear discrete-time systems," *International J. Control*, vol. 31, pp. 495–506, 1980.

Wonham, W. M., "On pole assignment in multi-input controllable linear systems," *IEEE Trans. Automatic Control*, vol. AC-12, pp. 660–665, 1967.

CHAPTER 4

Filters for Linear Stochastic Discrete-Time Systems

Observed variables in the real economy will be accompanied by observation errors, and any assumed linear dynamic economic system must contain a random disturbance term unexplainable by the system scheme. "Filtering" here is the problem of estimating all state variables from hitherto available input and output data in such a system. Filtering is, of course, a prerequisite for providing optimal feedback control values. Following Rhodes (1971), we verify the Kalman recursive formulas for the predictor and filter for our dynamic system in a tutorial manner (Section 4.2). (Section 4.1 is devoted to a preliminary least-squares estimation.) In Sections 4.3 and 4.4 we are concerned with the problem of finding a minimal-order filter of observer type, and we rely exclusively on Tse and Athans (1970). Their idea is to minimize the covariance matrix of estimation errors and to establish a dynamic minimal-order observer-estimator. The estimator is also applied to a general distributed-lag system. Finally, in Section 4.5, related economic applications are presented and examined.

4.1. Preliminary Least-Squares Estimators

Economic variables in any time sequence are always accompanied by disturbances not explainable by a linear system scheme. Taking this fact into consideration, we try to estimate variables in stochastic discrete-time systems. In particular, in this chapter we are concerned with estimating state vector x in the following linear discrete-time system composed of (1) and (2) and containing additive disturbances and observation errors. First,

$$x(t) = Ax(t-1) + Bv(t) + D\xi(t), \tag{1}$$

where ξ is the disturbance p vector, x and v are the state n vector and the control m vector, respectively, and A, B, D are constant matrices of appropriate dimensions. State vector x is accordingly stochastic, while control vector v is regarded as nonstochastic. (Vectors are assumed to be column ones.) Secondly, some state variables may not be directly observed. We therefore assume that r vector y with $r < n$ is the output through which state vector x can be imperfectly observed, being accompanied with an additive observation error:

$$y(t) = Cx(t) + \zeta(t), \tag{2}$$

where ζ stands for the r vector of observation errors and C is an $r \times n$ constant matrix of rank r. ξ and ζ are mutually independent stochastic vectors, not necessarily Gaussian, serially uncorrelated, with zero means and finite constant covariance matrices Ξ and Φ, respectively, viz.,

$$\begin{aligned} E\xi(t) = 0, \qquad E\zeta(t) &= 0 \\ \operatorname{cov}(\xi(t), \zeta(s)) &= 0 \\ \operatorname{cov}(\xi(t), \xi(s)) &= \Xi\delta_{ts} \\ \operatorname{cov}(\zeta(t), \zeta(s)) &= \Phi\delta_{ts} \end{aligned} \tag{3}$$

with $\delta_{tt} = 1$, $\delta_{ts} = 0$ $(t \neq s)$. Further, we assume that Ξ is positive semidefinite and that Φ is positive definite.

Kalman (1960) provided recursive formulas for optimal estimators for $x(t)$ of system (1), (2) under the assumption (3) with Gaussian stochastic disturbance and error terms.We shall verify the Kalman recursive formulas in the same system with non-Guassian disturbance and error terms, following Rhodes (1971). Optimal estimation means here that the expected value of the squared norm of linear estimation error is to be minimized, i.e., that, letting $\hat{x}$ be a linear estimator for x,

$$E(\|\tilde{x}\|^2) = E(\tilde{x}^T\tilde{x}) \tag{4}$$

is minimized for estimation error vector $\tilde{x}$ $(= x - \hat{x})$. Hence our estimators are linear least-squares estimators. We start by establishing (linear) least-squares estimation formulas.

Lemma 1 (Least-Squares Estimator When Only One Explaining Variable Vector Exists). *Let x, y be stochastic variable vectors such that the least-squares estimator $\hat{x}$ for x has a linear relation with explaining variable vector y:*

$$\hat{x} = Hy + b. \tag{5}$$

The matrix H and the vector b to minimize $E(\|\tilde{x}\|^2)$ for

$$\tilde{x} \equiv x - \hat{x} = x - Hy - b \tag{6}$$

are obtained as

$$H = S_{xy}S_{yy}^{-1}, \qquad b = m_x - Hm_y, \tag{7}$$

and hence

$$\hat{x} = S_{xy}S_{yy}^{-1}(y - m_y) + m_x \equiv E^*\{x \mid y\}, \tag{8}$$

where

$$\begin{aligned} m_x &\equiv Ex, \qquad m_y \equiv Ey, \\ S_{xy} &\equiv \operatorname{cov}(x, y) \\ &= E\{(x - m_x)(y - m_y)^T\} \\ &= E(xy^T) - m_x m_y^T, \end{aligned} \tag{9}$$

and $S_{yy} \equiv \operatorname{cov}(y, y)$ *is assumed to be positive definite.*

PROOF. Equation (4) can be expressed as

$$E(\|\tilde{x}\|^2) = E(\tilde{x}^T\tilde{x}) = \operatorname{tr}[E(\tilde{x}\tilde{x}^T)]. \tag{10}$$

Since

$$\begin{aligned} E(\tilde{x}\tilde{x}^T) &= E\{(x - Hy - b)(x - Hy - b)^T\} \\ &= E(xx^T) - HE(yx^T) - bm_x^T - E(xy^T)H^T + HE(yy^T)H^T \\ &\quad + bm_y^T H^T - m_x b^T + Hm_y b^T + bb^T \\ &= S_{xx} + m_x m_x^T - H(S_{yx} + m_y m_x^T) - bm_x^T - (S_{xy} + m_x m_y^T)H^T \\ &\quad + H(S_{yy} + m_y m_y^T)H^T + bm_y^T H^T - m_x b^T + Hm_y b^T + bb^T, \end{aligned}$$

(10) becomes

$$\begin{aligned} E(\|\tilde{x}\|^2) &= \operatorname{tr}(S_{xx} - HS_{yx} - S_{xy}H^T + HS_{yy}H^T) + b^T b - 2b^T m_x \\ &\quad + m_x^T m_x + 2m_y^T H^T(b - m_x) + m_y^T H^T H m_y \\ &= \operatorname{tr}[(S_{xx} - S_{xy}S_{yy}^{-1}S_{yx}) + (H - S_{xy}S_{yy}^{-1})S_{yy}(H^T - S_{yy}^{-1}S_{yx})] \\ &\quad + (b - m_x + Hm_y)^T(b - m_x + Hm_y). \end{aligned} \tag{11}$$

Define a norm for matrix $F \equiv H - S_{xy}S_{yy}^{-1}$ as

$$\|F\| \equiv [\operatorname{tr}(FS_{yy}F^T)]^{1/2}. \tag{12}$$

It can be seen that $\|F\|$ in (12) satisfies the following four axioms for matrix norm:

1) $\|F\| \geqslant 0$, and $\|F\| = 0$ if and only if $F = 0$,
2) $\|aF\| = |a|\,\|F\|$ for an arbitrary scalar a,
3) $\|F + G\| \leqslant \|F\| + \|G\|$,
4) $\|FG\| \leqslant \|F\|\,\|G\|$.

Thus (11) is rewritten

$$E(\|x\|^2) = \operatorname{tr}(S_{xx} - S_{xy}S_{yy}^{-1}S_{yx}) + \|H - S_{xy}S_{yy}^{-1}\|^2 + \|b - m_x + Hm_y\|^2,$$

from which (7) follows. □

Covariance matrices of the above estimation error $\tilde{x}$ are given by (15) below. Since

$$E\tilde{x} = Ex - E\hat{x} = S_{xy}S_{yy}^{-1}E(y - m_y) = 0, \tag{13}$$

(unbiasedness of $\hat{x}$), and since

$$\tilde{x} = (x - m_x) - S_{xy}S_{yy}^{-1}(y - m_y), \tag{14}$$

we have

$$\begin{aligned} \operatorname{cov}(\tilde{x}, \tilde{x}) &= E(\tilde{x}\tilde{x}^T) \\ &= S_{xx} - S_{xy}S_{yy}^{-1}S_{yx}. \end{aligned} \tag{15}$$

Similarly, by virtue of (13) and (14)

$$\operatorname{cov}(\tilde{x}, y) = E\big(\tilde{x}(y - m_y)^T\big) \equiv S_{xy} - S_{xy}S_{yy}^{-1}S_{yy} = 0. \tag{16}$$

Hence $\tilde{x}$ is uncorrelated with any linear function of y. In particular,

$$\operatorname{cov}(\tilde{x}, \hat{x}) = 0. \tag{17}$$

Lemma 2. *Let y be a stochastic vector, and let x, z be stochastic vectors of the same dimension, linearly dependent on y. Let M be a constant matrix, and c be a constant vector of the same dimension as x. Then using the notation in* (8), *we get*

$$E^*\{Mx + z + c \mid y\} = ME^*\{x \mid y\} + E^*\{z \mid y\} + c. \tag{18}$$

Proof. Defining

$$w \equiv Mx + z + c, \qquad m_w \equiv Mm_x + m_z + c, \tag{19}$$

we get from (8)

$$E^*\{w \mid y\} = S_{wy}S_{yy}^{-1}(y - m_y) + m_w,$$

into which we substitute

$$S_{wy} \equiv E\big((w - Mm_x - m_z - c)(y - m_y)^T\big) = MS_{xy} + S_{zy}. \qquad \square$$

Note that if we define for the w mentioned in (19)

$$\tilde{w} \equiv w - E^*\{w \mid y\} = M\tilde{x} + \tilde{z} \qquad \big(\tilde{z} \equiv z - E^*\{z \mid y\}\big),$$

then immediately we have

$$\operatorname{cov}(\tilde{w}, \tilde{w}) = M\operatorname{cov}(\tilde{x}, \tilde{x})M^T + M\operatorname{cov}(\tilde{x}, \tilde{z}) + \operatorname{cov}(\tilde{z}, \tilde{x})M^T + \operatorname{cov}(\tilde{z}, \tilde{z}). \tag{20}$$

Lemma 3 (Least-Squares Estimators When Two Explaining Variable Vectors Exist). *Let y and z be stochastic vectors of appropriate dimensions, and* *denote by* $E^*\{x \mid y,z\}$ *the least-squares estimator for* x *linearly dependent on* y *and* z.

a) *If* y *is uncorrelated with* z, *we have*

$$E^*\{x \mid y,z\} = E^*\{x \mid y\} + E^*\{\tilde{x}_y \mid z\}, \tag{21}$$

$$\text{cov}(\tilde{x}_{yz}, \tilde{x}_{yz}) = S_{xx} - S_{xy}S_{yy}^{-1}S_{yx} - S_{xz}S_{zz}^{-1}S_{zx}, \tag{22}$$

where $E^*\{x \mid y\}$ *is as in* (8) *and*

$$E^*\{\tilde{x}_y \mid z\} \equiv S_{xz}S_{zz}^{-1}(z - m_z), \tag{23}$$

$$\tilde{x}_y \equiv x - E^*\{x \mid y\}, \qquad \tilde{x}_{yz} \equiv x - E^*\{x \mid y,z\}. \tag{24}$$

b) *If* y *is correlated with* z, *we have*

$$E^*\{x \mid y,z\} = E^*\{x \mid y\} + E^*\{\tilde{x}_y \mid \tilde{z}_y\}, \tag{25}$$

$$\text{cov}(\tilde{x}_{yz}, \tilde{x}_{yz}) = S_{xx} - S_{xy}S_{yy}^{-1}S_{yx} - \tilde{S}_{xz}S_{zz}^{*}\tilde{S}_{zx} \tag{26}$$

$$= \text{cov}(\tilde{x}_y, \tilde{x}_y) - \text{cov}(\tilde{x}_y, \tilde{z}_y)\left[\text{cov}(\tilde{z}_y, \tilde{z}_y)\right]^{-1}\text{cov}(\tilde{z}_y, \tilde{x}_y), \tag{26'}$$

where

$$\begin{aligned} \tilde{z}_y &\equiv z - E^*\{z \mid y\}, \\ E^*\{\tilde{x}_y \mid \tilde{z}_y\} &\equiv \tilde{S}_{xz}S_{zz}^{*}\tilde{z}_y, \end{aligned} \tag{27}$$

$$S_{zz}^{*} \equiv \left(S_{zz} - S_{zy}S_{yy}^{-1}S_{yz}\right)^{-1} = \left[\text{cov}(\tilde{z}_y, \tilde{z}_y)\right]^{-1}, \tag{28}$$

$$\tilde{S}_{xz} \equiv S_{xz} - S_{xy}S_{yy}^{-1}S_{yz} = \text{cov}(\tilde{x}_y, \tilde{z}_y). \tag{29}$$

Furthermore, the estimator is unbiased, i.e.,

$$E\left(E^*\{x \mid y,z\}\right) = Ex, \tag{13'}$$

and

$$\text{cov}(\tilde{x}_{yz}, y) = \text{cov}(\tilde{x}_{yz}, z) = 0. \tag{30}$$

Proof.

a) Using the notation

$$p^T \equiv (y^T, z^T), \qquad m_p^T \equiv \left(m_y^T, m_z^T\right),$$

we have

$$E^*\{x \mid y,z\} = E^*\{x \mid p\} = S_{xp}S_{pp}^{-1}(p - m_p) + m_x, \tag{31}$$

into which we substitute

$$S_{xp} \equiv \text{cov}(x, p) = \left[\text{cov}(x, y), \text{cov}(x, z)\right] = \left[S_{xy}, S_{xz}\right],$$

$$S_{pp} \equiv \text{cov}(p, p) = \begin{pmatrix} \text{cov}(y, y) & 0 \\ 0 & \text{cov}(z, z) \end{pmatrix} = \begin{pmatrix} S_{yy} & 0 \\ 0 & S_{zz} \end{pmatrix},$$

entailing (21). In view of (15), we have

$$\operatorname{cov}(\tilde{x}_{yz}, \tilde{x}_{yz}) = S_{xx} - S_{xp} S_{pp}^{-1} S_{px},$$

into which we substitute S_{xp} and S_{pp} mentioned above, yielding (22).

b) In view of (16), $\tilde{z}_y$ is uncorrelated with y. Hence from (21)

$$E^*\{x \mid y, \tilde{z}_y\} = E^*\{x \mid y\} + E^*\{\tilde{x}_y \mid \tilde{z}_y\}. \tag{21'}$$

Also from (8), with $E\tilde{z}_y = E\tilde{x}_y = 0$ (cf. (13)) taken into account,

$$E^*\{\tilde{x}_y \mid \tilde{z}_y\} = \operatorname{cov}(\tilde{x}_y, \tilde{z}_y)\left[\operatorname{cov}(\tilde{z}_y, \tilde{z}_y)\right]^{-1} \tilde{z}_y. \tag{27'}$$

We now show (27′) to be equal to (27). In view of (15),

$$\operatorname{cov}(\tilde{z}_y, \tilde{z}_y) = S_{zz} - S_{zy} S_{yy}^{-1} S_{yz}. \tag{15'a}$$

$$\begin{aligned} \operatorname{cov}(\tilde{x}_y, \tilde{z}_y) &= E\left(\tilde{x}_y \tilde{z}_y^T\right) \\ &= E\left[(x - E^*\{x \mid y\})(z - E^*\{z \mid y\})^T\right] \\ &= E\Big[\left((x - m_x) - S_{xy} S_{yy}^{-1}(y - m_y)\right) \\ &\qquad \times \left((z - m_z) - S_{zy} S_{yy}^{-1}(y - m_y)\right)^T\Big], \end{aligned} \tag{15'b}$$

which is easily seen to be equal to (29). Next we verify that

$$E^*\{x \mid y, z\} = E^*\{x \mid y, \tilde{z}_y\}.$$

The left-hand side of the above is expressed as (31), into which

$$S_{xp} = \left[S_{xy}, S_{xz}\right], \quad \text{and}$$

$$S_{pp}^{-1} = \begin{bmatrix} S_{yy} & S_{yz} \\ S_{zy} & S_{zz} \end{bmatrix}^{-1} = \begin{bmatrix} S_{yy}^{-1} + S_{yy}^{-1} S_{yz} S_{zz}^* S_{zy} S_{yy}^{-1} & -S_{yy}^{-1} S_{yz} S_{zz}^* \\ -S_{zz}^* S_{zy} S_{yy}^{-1} & S_{zz}^* \end{bmatrix}$$

are substituted to result in

$$\begin{aligned} E^*\{x \mid y, z\} &= S_{xy} S_{yy}^{-1}(y - m_y) + m_x + \tilde{S}_{xx} S_{zz}^*\left((z - m_z) - S_{zy} S_{yy}^{-1}(y - m_y)\right) \\ &= E^*\{x \mid y\} + \tilde{S}_{xz} S_{zz}^*(z - E^*\{x \mid z\}) \end{aligned}$$

since $S_{xp} S_{pp}^{-1} = [S_{xy} S_{yy}^{-1} - \tilde{S}_{xz} S_{zz}^* S_{zy} S_{yy}^{-1}, \tilde{S}_{xz} S_{zz}^*]$.

Equation (26) is proved as follows.

$$\begin{aligned} \operatorname{cov}(\tilde{x}_{yz}, \tilde{x}_{yz}) &= E\left(\tilde{x}_{yz} \tilde{x}_{yz}^T\right) \\ &= E\left[(x - E^*\{x \mid y, z\})(x - E^*\{x \mid y, z\})^T\right], \end{aligned}$$

in which we take account of

$$\begin{aligned} x - E^*\{x \mid y, z\} = (x - m_x) &- (S_{xy} - \tilde{S}_{xz} S_{zz}^* S_{zy}) S_{yy}^{-1}(y - m_y) \\ &- \tilde{S}_{xz} S_{zz}^*(z - m_z), \end{aligned}$$

to entail

$$\begin{aligned}
&\operatorname{cov}(\tilde{x}_{yz}, \tilde{x}_{yz}) \\
&\quad = S_{xx} - (S_{xy} - \tilde{S}_{xz} S_{zz}^{*} S_{zy}) S_{yy}^{-1} S_{yx} - \tilde{S}_{xz} S_{zz}^{*} S_{zx} \\
&\qquad + \left[- S_{xy} + S_{xy} - \tilde{S}_{xz} S_{zz}^{*} S_{zy} + \tilde{S}_{xz} S_{zz}^{*} S_{zy} \right] S_{yy}^{-1} (S_{yx} - S_{yz} S_{zz}^{*} \tilde{S}_{zx}) \\
&\qquad + \left[- S_{xz} + (S_{xy} - \tilde{S}_{xz} S_{zz}^{*} S_{zy}) S_{yy}^{-1} S_{yz} + \tilde{S}_{xz} S_{zz}^{*} S_{zz} \right] S_{zz}^{*} \tilde{S}_{zx} .
\end{aligned}$$

The two square-bracketed terms on the right-hand side of this equation vanish in view of (28) and (29).

Last we verify (30). Since

$$\begin{aligned}
E(\tilde{x}_{yz}) &= Ex - E(E^{*}\{ x \mid y, z \}) \\
&= E(x - E^{*}\{ x \mid y \}) - E(E^{*}\{ \tilde{x}_{y} \mid \tilde{z}_{y} \}) = 0
\end{aligned}$$

(unbiasedness of $E^{*}\{ x \mid y, z \}$), we get

$$\begin{aligned}
\operatorname{cov}(\tilde{x}_{yz}, y) &= E\left(\tilde{x}_{yz} (y - m_{y})^{T}\right) \\
&= E\left[(x - m_{x})(y - m_{y})^{T} \right] \\
&\quad - E\Big[S_{xy} S_{yy}^{-1} (y - m_{y})(y - m_{y})^{T} \\
&\qquad + \tilde{S}_{xz} S_{zz}^{*} \big((z - m_{z}) - S_{zy} S_{yy}^{-1} (y - m_{y}) \big) (y - m_{y})^{T} \Big] \\
&= S_{xy} - S_{xy} - \tilde{S}_{xz} S_{zz}^{*} (S_{zy} - S_{zy}) \\
&= 0.
\end{aligned}$$

Proceed similarly for $\operatorname{cov}(\tilde{x}_{yz}, z) = 0$. □

Lemma 3 can be generalized to the case where arbitrarily many successively occurring vectors exist.

Theorem 1 (Rhodes, 1971). *Let $y(1), y(2), \ldots, y(t)$ be t successively occurring vectors, and let $\hat{x}(t+1 \mid t)$ denote the least-squares estimator of $x(t+1)$ linearly dependent on these vectors, i.e.,*

$$\hat{x}(t+1 \mid t) \equiv E^{*}\{ x(t+1) \mid y(1), y(2), \ldots, y(t) \}.$$

Then the estimator is expressed in a recursive form

$$\begin{aligned}
\hat{x}(t+1 \mid t) = \hat{x}(t+1 \mid t-1) &+ \operatorname{cov}(\tilde{x}(t+1 \mid t-1), \tilde{y}(t \mid t-1)) \\
&\times \left[\operatorname{cov}(\tilde{y}(t \mid t-1), \tilde{y}(t \mid t-1)) \right]^{-1} \tilde{y}(t \mid t-1), \qquad (32)
\end{aligned}$$

where

$$\begin{aligned}
\tilde{x}(t+1 \mid t-1) &\equiv x(t+1) - \hat{x}(t+1 \mid t-1), \\
\tilde{y}(t \mid t-1) &\equiv y(t) - \hat{y}(t \mid t-1), \\
\hat{y}(t \mid t-1) &\equiv E^{*}\{ y(t) \mid y(1), \ldots, y(t-1) \}.
\end{aligned}$$

The estimator is unbiased, i.e.,

$$E\hat{x}(t+1 \mid t) = Ex(t+1). \tag{33}$$

Furthermore,

$$\begin{aligned}
&\operatorname{cov}(\tilde{x}(t+1 \mid t), \tilde{x}(t+1 \mid t)) \\
&\quad = \operatorname{cov}(\tilde{x}(t+1 \mid t-1), \tilde{x}(t+1 \mid t-1)) \\
&\qquad - \operatorname{cov}(\tilde{x}(t+1 \mid t-1), \tilde{y}(t \mid t-1))\left[\operatorname{cov}(\tilde{y}(t \mid t-1), \tilde{y}(t \mid t-1))\right]^{-1} \\
&\qquad \times \operatorname{cov}(\tilde{y}(t \mid t-1), \tilde{x}(t+1 \mid t-1)),
\end{aligned} \tag{34}$$

and

$$\operatorname{cov}(\tilde{x}(t+1 \mid t), y(s)) = 0 \qquad \textit{for} \quad s = 1, 2, \ldots, t. \tag{35}$$

Proof. Define a column vector

$$Y(t-1) \equiv \left(y^T(1), y^T(2), \ldots, y^T(t-1)\right)^T, \tag{1*}$$

and put

$$x = x(t+1), \qquad y = Y(t-1), \qquad z = y(t)$$

in equations (25)–(30). □

Theorem 1 is almost sufficient to deduce the Kalman filter. Here we note that the filter will be a least-squares estimator depending on the presumption that additive errors are simply white noises, as shown by (3), and not necessarily Gaussian. For the case of Gaussian white noise, the reader may refer to standard textbooks such as Meditch (1969), Jaswinski (1970), and Anderson and Moore (1979). A similar approach to our treatment in the following section is found in Bertsekas (1976, pp. 158–174).

4.2. Kalman Predictor and Filter

The so-called Kalman predictor or filter will be derived as a straightforward extension of the least-squares estimator in Theorem 1. We provide a lemma for a slight alteration to the estimator expression.

Lemma 4 (Matrix Inversion Lemma). *Let A, B, C, and D be matrices of dimensions $m \times m$, $n \times n$, $m \times n$, and $n \times m$, respectively. Assuming A and B to be nonsingular, we have*

$$(A^{-1} + DB^{-1}C)^{-1} = A - AD(B + CAD)^{-1}CA.$$

(The proof can be done by a direct multiplication:

$$(A^{-1} + DB^{-1}C)\left(A - AD(B + CAD)^{-1}CA\right) = I,$$

or by comparing (31) and (32) in Murata (1977, Section 1.1).)

We are now in a position to obtain Kalman's recursive formulas of optimal estimators.

Theorem 2 (Kalman Predictor). *Let $\hat{x}(t+1\mid t)$ denote the linear least-squares estimator for $x(t+1)$ in terms of $y(s)$, $s=0,1,2,\ldots,t$, for system* (1), (2) *under the assumption* (3), *and let $\tilde{x}(t+1\mid t)$ be its estimation error, i.e.,*

$$\tilde{x}(t+1\mid t)\equiv x(t+1)-\hat{x}(t+1\mid t). \tag{36}$$

The estimator, termed the Kalman perdictor, is expressed in a recursive form

$$\hat{x}(t+1\mid t)=A\big[I-N(t)C\big]\hat{x}(t\mid t-1)+AN(t)y(t)+Bv(t+1), \tag{37}$$

where

$$N(t)\equiv S(t)C^T\big[CS(t)C^T+\Phi\big]^{-1}, \tag{38}$$

$$S(t)\equiv \operatorname{cov}(\tilde{x}(t\mid t-1),\tilde{x}(t\mid t-1)), \tag{39}$$

and $S(t)$ is assumed to be positive definite for all t. The estimator is unbiased, i.e.,

$$E\hat{x}(t+1\mid t)=Ex(t+1). \tag{33}$$

Furthermore, error covariance matrices are computable in a recursive manner by the following Riccati difference equation (cf. (28) of Section 2.2)

$$S(t+1)=AS(t)\Big[I-C^T\big(CS(t)C^T+\Phi\big)^{-1}CS(t)\Big]A^T+D\Xi D^T \tag{40}$$

$$=A\big[S(t)^{-1}+C^T\Phi^{-1}C\big]^{-1}A^T+D\Xi D^T. \tag{41}$$

The initial values for the recursive computation are

$$\hat{x}(0\mid -1)=Ex(0), \tag{42a}$$

$$S(0)=\operatorname{cov}(x(0),x(0)). \tag{42b}$$

Proof. We apply Theorem 1 here. By an iterative substitution, (1) is transformed into

$$x(t)=A^{t-s}x(s)+\sum_{k=s+1}^{t}A^{t-k}(Bv(k)+D\xi(k)). \tag{2*}$$

Using notation (1*) and taking Lemma 2 into consideration, we get

$$\begin{aligned}\hat{x}(t+1\mid t-1)&=E^*\{x(t+1)\mid Y(t-1)\}\\&=E^*\{Ax(t)+Bv(t+1)+D\xi(t+1)\mid Y(t-1)\}\\&=AE^*\{x(t)\mid Y(t-1)\}+Bv(t+1)\end{aligned} \tag{3*}$$

In deriving (3*), we have taken account of

$$E^*\{\xi(t+1)\mid Y(t-1)\}=0 \tag{4*}$$

in view of (8), $E\xi(t+1)=0$, and $\operatorname{cov}(\xi(t+1),Y(t-1))=0$. Equation (3*) is rewritten

$$\hat{x}(t+1 \mid t-1) = A\hat{x}(t \mid t-1) + Bv(t+1). \tag{43}$$

Hence by definition (36)

$$\begin{aligned}\tilde{x}(t+1 \mid t-1) &= Ax(t) + Bv(t+1) + D\xi(t+1) \\ &\quad - A\hat{x}(t \mid t-1) - Bv(t+1) \\ &= A\tilde{x}(t \mid t-1) + D\xi(t+1). \end{aligned} \tag{43'}$$

Similarly,

$$\begin{aligned}\hat{y}(t \mid t-1) &\equiv E^*\{y(t) \mid Y(t-1)\} \\ &= E^*\{Cx(t) + \zeta(t) \mid Y(t-1)\} \\ &= C\hat{x}(t \mid t-1). \qquad (E\zeta(t) = 0, \operatorname{cov}(\zeta(t), Y(t-1)) = 0) \end{aligned} \tag{44}$$

$$\begin{aligned}\tilde{y}(t \mid t-1) &\equiv y(t) - \hat{y}(t \mid t-1) \\ &= C\tilde{x}(t \mid t-1) + \zeta(t). \end{aligned} \tag{44'}$$

Considering (44′), (2*), and taking $\operatorname{cov}(x(t), \zeta(s)) = 0$,

$$\operatorname{cov}(v(k), \tilde{x}(s \mid s-1)) = \operatorname{cov}(\xi(k), \tilde{x}(s \mid s-1)) = 0 \qquad \text{for} \quad k > s$$

into account, we calculate

$$\begin{aligned}&\operatorname{cov}(x(t), \tilde{y}(s \mid s-1)) \\ &\quad = \operatorname{cov}[x(t), \tilde{x}(s \mid s-1)]C^T \\ &\quad = \operatorname{cov}\left[A^{t-s}x(s) + \sum_{k=s+1}^{t} A^{t-k}(Bv(k) + D\xi(k)), \tilde{x}(s \mid s-1)\right]C^T \\ &\quad = A^{t-s}\operatorname{cov}[x(s), \tilde{x}(s \mid s-1)]C^T \\ &\quad = A^{t-s}\operatorname{cov}[\tilde{x}(s \mid s-1) + \hat{x}(s \mid s-1), \tilde{x}(s \mid s-1)]C^T \\ &\quad = A^{t-s}\operatorname{cov}[\tilde{x}(s \mid s-1), \tilde{x}(s \mid s-1)]C^T \qquad (\text{cf. } (17)) \\ &\quad = A^{t-s}S(s)C^T \qquad \text{for} \quad s < t. \end{aligned} \tag{45}$$

As $\hat{x}(t \mid s-1)$ is uncorrelated with $\tilde{x}(t \mid s-1)$ and $\zeta(s)$, so is it with $\tilde{y}(t \mid s-1)$ in view of (44′). Thus from (45) it follows that

$$\begin{aligned}\operatorname{cov}(\tilde{x}(t \mid s-1), \tilde{y}(s \mid s-1)) &= \operatorname{cov}(x(t) - \hat{x}(t \mid s-1), \tilde{y}(s \mid s-1)) \\ &= A^{t-s}S(s)C^T \qquad \text{for} \quad s < t. \end{aligned} \tag{45'}$$

Also, as $\zeta(t)$ is uncorrelated with $x(t)$ and $\hat{x}(t \mid t-1)$, so is it with $\tilde{x}(t \mid t-1)$. Therefore, considering (44′), we get

$$\operatorname{cov}(\tilde{y}(t \mid t-1), \tilde{y}(t \mid t-1)) = CS(t)C^T + \Phi. \tag{46}$$

Substitution of (43), (45′), (46), and (44) into (32) yields (37).

Next, in order to prove (40), it suffices to substitute (45′), (46) and

$$\operatorname{cov}\left[\tilde{x}(t+1 \mid t-1), \tilde{x}(t+1 \mid t-1)\right] = A \operatorname{cov}\left[\tilde{x}(t \mid t-1), \tilde{x}(t \mid t-1)\right] A^T + D \Xi D^T, \tag{47}$$

into (34). Equation (47) is obtained by considering (43′) and noting that $\xi(t+1)$ is uncorrelated with $Y(t-1)$, hence with $\hat{x}(t \mid t-1)$, and that $\xi(t+1)$ is uncorrelated with $x(t)$.

Equation (41) is derived by applying the matrix inversion lemma (Lemma 4) to (40).

Last, (42a) and (42b) follow from (8) and (15), respectively, since $y(s)$ does not exist for $s < 0$. □

The estimator (37) is the predictor of $x(t+1)$ based on information in period t, while the so-called Kalman filter is intended to be the least-squares estimator for $x(t)$ based on the contemporaneous information.

Theorem 3 (Kalman Filter). *Let $\hat{x}(t \mid t)$ denote the linear least-squares estimator for $x(t)$ in terms of $y(s)$, $s = 0, 1, 2, \ldots, t$, for system* (1), (2) *under the assumption* (3), *and let $\tilde{x}(t \mid t)$ be its estimation error, i.e.,*

$$\tilde{x}(t \mid t) \equiv x(t) - \hat{x}(t \mid t). \tag{36′}$$

The estimator, termed the Kalman filter, is expressed in a recursive form

$$\hat{x}(t \mid t) = \left[I - N(t)C\right]\left(A\hat{x}(t-1 \mid t-1) + Bv(t)\right) + N(t)y(t), \tag{48}$$

where $N(t)$ is that of (38) *and the corresponding $S(t)$ is found from the recursive relation* (40) *or* (41). *The estimator is unbiased, i.e.,*

$$E\hat{x}(t \mid t) = Ex(t). \tag{33′}$$

Defining the covariance matrix of the estimation error as $\bar{S}(t) \equiv \operatorname{cov}(\tilde{x}(t \mid t), \tilde{x}(t \mid t))$, we get the recursive expression

$$\bar{S}(t) = S(t) - S(t)C^T\left[CS(t)C^T + \Phi\right]^{-1}CS(t), \tag{49}$$

or

$$\bar{S}(t) = \left[\left(A\bar{S}(t-1)A^T + D\Xi D^T\right)^{-1} + C^T\Phi^{-1}C\right]^{-1} \tag{49′}$$

provided $S(t)$ is positive definite for all t. The initial values for the recursive computation are

$$\hat{x}(0 \mid 0) = Ex(0), \tag{42′a}$$

$$\bar{S}(0) = \left[S(0)^{-1} + C^T\Phi^{-1}C\right]^{-1}, \tag{42′b}$$

where $S(0)$ is that of (42b).

PROOF. We again apply Theorem 1. Using notation (1*) and taking account of Lemma 2, and $E^*\{\xi(t)\mid Y(t-1)\}=0$ by virtue of $E\xi(t)=0$ and $\operatorname{cov}(\xi(t), Y(t-1))=0$, we obtain

$$\begin{aligned}\hat{x}(t\mid t-1) &\equiv E^*\{x(t)\mid Y(t-1)\}\\ &= E^*\{Ax(t-1)+Bv(t)+D\xi(t)\mid Y(t-1)\}\\ &= A\hat{x}(t-1\mid t-1)+Bv(t),\end{aligned}\tag{50}$$

and then

$$\begin{aligned}\tilde{x}(t\mid t-1) &\equiv x(t)-\hat{x}(t\mid t-1)\\ &= Ax(t-1)+Bv(t)+D\xi(t)-A\hat{x}(t-1\mid t-1)-Bv(t)\\ &= A\tilde{x}(t-1\mid t-1)+D\xi(t).\end{aligned}\tag{51}$$

Expressions (32) and (34) with time delay by one period in $\hat{x}$ and $\tilde{x}$ are

$$\begin{aligned}\hat{x}(t\mid t) = \hat{x}(t\mid t-1) &+ \operatorname{cov}(\tilde{x}(t\mid t-1), \tilde{y}(t\mid t-1))\\ &\times\left[\operatorname{cov}(\tilde{y}(t\mid t-1), \tilde{y}(t\mid t-1))\right]^{-1}\tilde{y}(t\mid t-1),\end{aligned}\tag{32'}$$

and

$$\begin{aligned}\operatorname{cov}(\tilde{x}(t\mid t),\tilde{x}(t\mid t)) = \operatorname{cov}(\tilde{x}(t\mid t-1),\tilde{x}(t\mid t-1))&\\ -\operatorname{cov}(\tilde{x}(t\mid t-1), \tilde{y}(t\mid t-1))&\\ \times\left[\operatorname{cov}(\tilde{y}(t\mid t-1), \tilde{y}(t\mid t-1))\right]^{-1}&\\ \times\operatorname{cov}(\tilde{y}(t\mid t-1),\tilde{x}(t\mid t-1)).&\end{aligned}\tag{34'}$$

First, since $\xi(t)$ is uncorrelated with $\tilde{x}(t-1\mid t-1)$,

$$\begin{aligned}S(t) &\equiv \operatorname{cov}(\tilde{x}(t\mid t-1),\tilde{x}(t\mid t-1))\\ &= A\bar{S}(t-1)A^T+D\Xi D^T.\end{aligned}\tag{40'}$$

Next for $s<t$, in view of $\operatorname{cov}(\zeta(s+1),x(t))=0$, (2*), (36′), and (17), we have

$$\begin{aligned}\operatorname{cov}\left[x(t), \tilde{y}(s+1\mid s)\right] &= \operatorname{cov}\left[x(t),\tilde{x}(s+1\mid s)\right]C^T\\ &= A^{t-s-1}\operatorname{cov}\left[Ax(s)+Bv(s+1)\right.\\ &\qquad\left.+D\xi(s+1),A\tilde{x}(s\mid s)+D\xi(s+1)\right]C^T\\ &= A^{t-s}\operatorname{cov}\left[x(s),\tilde{x}(s\mid s)\right]A^TC^T+A^{t-s-1}D\Xi D^TC^T\\ &= A^{t-s-1}\left(A\bar{S}(s)A^T+D\Xi D\right)^TC^T,\end{aligned}$$

and hence referring to (45′), we get

$$\begin{aligned}\operatorname{cov}\left[\tilde{x}(t\mid s), \tilde{y}(s+1\mid s)\right] &= \operatorname{cov}\left[x(t)-\hat{x}(t\mid s), \tilde{y}(s+1\mid s)\right]\\ &= A^{t-s-1}\left(A\bar{S}(s)A^T+D\Xi D^T\right)C^T.\end{aligned}\tag{52}$$

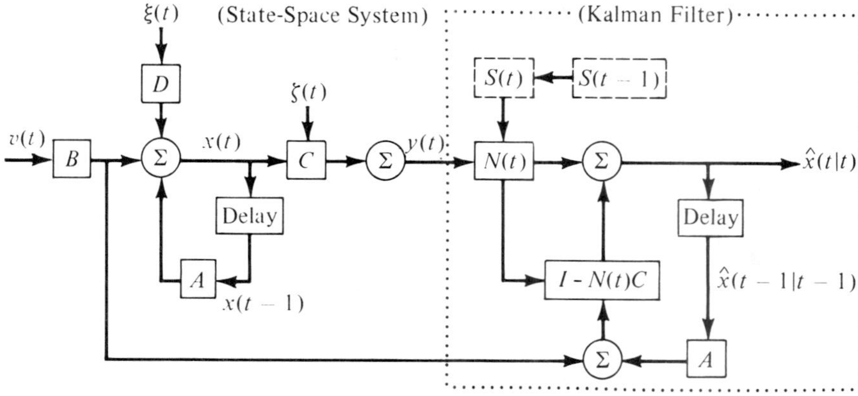

Figure 2. Kalman Filter Connected with State-Space System
($\sum$ is summation, and Delay means one-period delay.)

Putting $s = t - 1$ in (52) yields

$$\operatorname{cov}\left[\tilde{x}(t \mid t-1), \tilde{y}(t \mid t-1)\right] = \left(A\bar{S}(t-1)A^T + D\Xi D^T\right)C^T. \quad (52')$$

Substituting (50), (52′), (46), and (44) into (33′), with (40′) taken into account, results in (48). A similar substitution in (34′) yields (49), and application of the matrix inversion lemma (Lemma 4) to (40) gives

$$\bar{S}(t) = \left[S(t)^{-1} + C^T\Phi^{-1}C\right]^{-1}, \quad (49'')$$

which is nothing but (49′). Equation (42′b) follows at once from (49″). □

Remark. Kalman predictors and filters are equal to the so-called minimum variance unbiased estimators in view of (4) and unbiasedness of the estimators (cf. (33) and (33′)).

Kalman filter $\hat{x}(t \mid t)$ in (48) is schematized in a flow chart (Fig. 2), in connection with state-space system (1), (2). Here we note that an efficient algorithm is proposed by Morf *et al.* (1974) for Kalman predictor. Next we verify the stability property of Kalman predictor.

Theorem 4 (Stability of Kalman Predictor). *If the system*

$$x(t) = Ax(t-1) + D\xi(t) \quad (1')$$

$$y(t) = Cx(t) \quad (2')$$

is state controllable and observable, and if Ξ, Φ are positive definite, then the coefficient matrix $A[I - N(t)C]$ in (37) tends asymptotically to a stable matrix as planning horizon is extended.

Proof. [All reference theorems in this proof are from Murata (1977, Section 9.4).] System (1′), (2′) is state controllable and observable if and only if the following dual system (53), (54) is observable and state controllable (cf. Theorem 18):

$$z(t) = -A^T z(t-1) - C^T \eta(t), \tag{53}$$

$$w(t) = D^T z(t). \tag{54}$$

Given the final-time boundary condition $z(\beta)$, we define a cost function J over the periods from β to 0 backward in time as

$$J \equiv \sum_{t=\beta}^{1} \{\eta^T(t)\Phi\eta(t) + w^T(t)\Xi w(t)\} + z^T(0)S(0)z(0), \tag{55}$$

and optimize $\eta(t)$ such that J is minimized for system (53), (54). Then, by virtue of an analogy to Theorem 20′, a unique $S(t)$ exists satisfying the Riccati difference equation (40). Now, extend the initial period back to $-\infty$, and minimize a modified cost

$$J \equiv \sum_{t=\beta}^{-\infty} \{\eta^T(t)\Phi\eta(t) + w^T(t)\Xi w(t)\} \tag{55′}$$

subject to system (53), (54). Then, by Theorem 21′, the solution $S(t)$ of (40) becomes S given by

$$S = AS[I - NC]^T A^T + D\Xi D^T, \tag{40″}$$

where

$$N \equiv SC^T(CSC^T + \Phi)^{-1}, \tag{38′}$$

and each eigenvalue of $A[I - NC]$ is less than unity in modulus, i.e., $A[I - N(t)C]$ approches asymptotically to a stable matrix $A[I - NC]$ as planning horizon stretches. □

For further studies of the asymptotic stability, see Deyst and Price (1968). Concluding this section, we give an application of Kalman filter to a macroeconomic model.

Application 1. Our model consists of three equations (5*), (6*), and (7*), representing money market, the central bank's behavior, and the goods market, respectively:

$$r(t) = \beta_1 m(t-1) + \beta_2 q(t-1) + \beta_3 s(t) + \xi_1(t) \tag{5*}$$

$$m(t) = \beta_4 r(t-1) + \beta_5 p(t) + \xi_2(t) \tag{6*}$$

$$q(t) = \beta_6 r(t-1) + \xi_3(t), \tag{7*}$$

where r is the nominal rate of interest, m is the banking sector's stock of deposit liabilities, q is the nominal GNP, s is the additional supply of money, p is the banking sector's desired stock of reserves, and ξ_i $(i = 1, 2, 3)$ stand for mutually independent, serially uncorrelated disturbances having zero means and finite variances:

$$E\xi_i(t) = 0, \qquad i = 1, 2, 3 \tag{8*}$$

$$E\xi_i^2(t) = \Xi_i < \infty, \qquad i = 1, 2, 3. \tag{9*}$$

Among the variables, s and p are supposed to be control variables (instruments) and the others are state variables. Thus the state-space representation of the above model is

$$\begin{bmatrix} r(t) \\ m(t) \\ q(t) \end{bmatrix} = \begin{bmatrix} 0 & \beta_1 & \beta_2 \\ \beta_4 & 0 & 0 \\ \beta_6 & 0 & 0 \end{bmatrix} \begin{bmatrix} r(t-1) \\ m(t-1) \\ q(t-1) \end{bmatrix} + \begin{bmatrix} \beta_3 & 0 \\ 0 & \beta_5 \\ 0 & 0 \end{bmatrix} \begin{pmatrix} s(t) \\ p(t) \end{pmatrix} + \begin{bmatrix} \xi_1(t) \\ \xi_2(t) \\ \xi_3(t) \end{bmatrix}. \tag{10*}$$

We assume that $r(t)$ and $m(t)$ are observed with observation errors $\zeta_1(t)$ and $\zeta_2(t)$, respectively, while $q(t)$ is not directly observed contemporaneously. Denoting by $\bar{r}$ and $\bar{m}$ the observed values of r and m, therefore, we have output equation

$$\begin{pmatrix} \bar{r}(t) \\ \bar{m}(t) \end{pmatrix} = \begin{pmatrix} 1 & 0 & 0 \\ 0 & 1 & 0 \end{pmatrix} \begin{bmatrix} r(t) \\ m(t) \\ q(t) \end{bmatrix} + \begin{pmatrix} \zeta_1(t) \\ \zeta_2(t) \end{pmatrix}, \tag{11*}$$

with

$$E\zeta_i(t) = 0, \qquad E\zeta_i^2(t) = \Phi_i < \infty, \qquad i = 1, 2. \tag{12*}$$

Equations (10*) and (11*) correspond to (1) and (2), respectively, with the following notations:

$$x(t) \equiv \begin{bmatrix} r(t) \\ m(t) \\ q(t) \end{bmatrix}, \quad A \equiv \begin{bmatrix} 0 & \beta_1 & \beta_2 \\ \beta_4 & 0 & 0 \\ \beta_6 & 0 & 0 \end{bmatrix}, \quad B \equiv \begin{bmatrix} \beta_3 & 0 \\ 0 & \beta_5 \\ 0 & 0 \end{bmatrix}, \quad \xi(t) \equiv \begin{bmatrix} \xi_1(t) \\ \xi_2(t) \\ \xi_3(t) \end{bmatrix},$$

$$v(t) \equiv \begin{pmatrix} s(t) \\ p(t) \end{pmatrix}, \quad y(t) \equiv \begin{pmatrix} \bar{r}(t) \\ \bar{m}(t) \end{pmatrix}, \quad C \equiv \begin{pmatrix} 1 & 0 & 0 \\ 0 & 1 & 0 \end{pmatrix}, \quad \zeta(t) \equiv \begin{pmatrix} \zeta_1(t) \\ \zeta_2(t) \end{pmatrix},$$

and $D = I$ (identity matrix). We also denote

$$\Xi \equiv \begin{bmatrix} \Xi_1 & 0 & 0 \\ 0 & \Xi_2 & 0 \\ 0 & 0 & \Xi_3 \end{bmatrix}, \qquad \Phi \equiv \begin{pmatrix} \Phi_1 & 0 \\ 0 & \Phi_2 \end{pmatrix}.$$

We may apply Theorem 3 (Kalman filter) to obtain $\hat{x}(t \mid t)$ which contains the estimator for $q(t)$ as well as the estimators for $r(t)$ and $m(t)$.

Thus, on the assumption that covariances among estimation errors $\tilde{r}(t \mid t-1)$, $\tilde{m}(t \mid t-1)$ and $\tilde{q}(t \mid t-1)$ are zeros with σ_1 and σ_2 representing the variances of $\tilde{r}(t \mid t-1)$ and $\tilde{m}(t \mid t-1)$ respectively, we get

$$\begin{aligned}
\hat{r}(t \mid t) &= (1-\lambda_1)\{\beta_1 \hat{m}(t-1 \mid t-1) \\
&\qquad + \beta_2 \hat{q}(t-1 \mid t-1) + \beta_3 s(t)\} + \lambda_1 \bar{r}(t), \\
\hat{m}(t \mid t) &= (1-\lambda_2)\{\beta_4 \hat{r}(t-1 \mid t-1) + \beta_5 p(t)\} + \lambda_2 \overline{m}(t), \\
\hat{q}(t \mid t) &= \beta_6 \hat{q}(t-1 \mid t-1),
\end{aligned}$$

where $\lambda_i \equiv \sigma_i/(\sigma_i + \Phi_i)$ for $i = 1, 2$.

If we employ $\bar{r}(t)$ and $\overline{m}(t)$, instead of the Kalman filters of $r(t)$ and $m(t)$, for feedback optimal control rule, then the only estimator needed will be that of $q(t)$ which is by no means directly measurable.

(For other applications of Kalman filter to economic problems, see Section 4.5.) Our aim in the next section is to estimate only those variables which are not directly accessible.

4.3. Minimal-Order Observer-Estimators: Existence

Kalman estimators are designed to estimate all state variables, while our minimal-order observer-estimators are intended to estimate only those state variables which are not directly observable. Suppose that output vector $y(t)$ in equation (2) corresponds to some r components of n vector $x(t)$. We shall establish our recursive formulas to estimate the rest $n-r$ variables, following Tse and Athans (1970). These formulas have a close relationship to the recursive minimum-cost observer (Section 3.4.).

We continue to consider system (1), (2) under assumption (3). Besides we assume that for any given initial value x_0 of $x(0)$, $x_0 - x(0)$ is uncorrelated with disturbances $\xi(t)$ and observation error $\zeta(t)$, and that

$$E(x_0 - x(0)) = 0. \tag{56}$$

Hence, denoting transposition by superscript T,

$$E\left[(x_0 - x(0))\xi^T(t)\right] = \operatorname{cov}(x_0 - x(0), \xi(t)) = 0, \tag{57}$$

$$E\left[(x_0 - x(0))\zeta^T(t)\right] = \operatorname{cov}(x_0 - x(0), \zeta(t)) = 0. \tag{57'}$$

Definition 1. Let C be an $r \times n$ real constant matrix with rank r and let T denote any $s \times n$ real matrix with $s = n - r$. The set

$$\Upsilon(C) = \{T : N(T) \cap N(C) = \emptyset\} \qquad (\emptyset \text{ denotes an empty set})$$

is called the set of complementary matrices of order $n-r$ for C, where $N(C)$ denotes the null space of C.

By virtue of this definition, the only vector b satisfying both $Cb = 0$ and $Tb = 0$ is a null vector, i.e., $n \times n$ matrix $[C^T, T^T]^T$ is nonsingular for each

$T \in \Upsilon(C)$ in view of Theorem 27 in Murata (1977, Section 2.3). Thus we have an $n \times r$ matrix $Q(t)$ and an $n \times s$ matrix $P(t)$ such that

$$Q(t)C + P(t)T(t) = I \qquad \text{for} \quad T(t) \in \Upsilon(C), \tag{58}$$

where $[Q(t), P(t)]$ is the inverse of $[C^T, T^T(t)]^T$. For an estimator $\hat{x}(t)$ for $x(t)$, define an order s observer-estimator as

$$z(t) \equiv T(t)\hat{x}(t) \tag{59}$$

and consider

$$z(t) = F(t)z(t-1) + H(t)y(t-1) + T(t)Bv(t) \tag{60}$$

for $t = 1, 2, \ldots$. (Note that the time structure of (60) is the same as of the Kalman predictor (37).) Substituting

$$F(t) = T(t)AP(t-1), \tag{61a}$$

$$H(t) = T(t)AQ(t-1). \tag{61b}$$

and (2) into (60) and subtracting $T(t)x(t)$ from the resultant equation, with (1) and (58) taken into account, we get

$$e(t) = T(t)AP(t-1)e(t-1) + T(t)AQ(t-1)\zeta(t-1) - T(t)D\xi(t), \tag{62}$$

for $t = 1, 2, \ldots,$ where $e(t)$ is an s-dimensional error:

$$e(t) \equiv z(t) - T(t)x(t). \tag{63}$$

The covariance matrix of $e(t)$, denoted $S(t)$, is assumed to be positive semidefinite. Define an n-dimensional error

$$\tilde{e}(t) \equiv P(t)e(t). \tag{63$'$}$$

Premultiplication of (62) by $P(t)$ and consideration of (58) yield

$$\tilde{e}(t) = [I - Q(t)C]\{A\tilde{e}(t-1) + AQ(t-1)\zeta(t-1) - D\xi(t)\} \tag{62$'$}$$

for $t = 1, 2, \ldots$. Choosing, in view of (59),

$$z(0) = T(0)x_0, \tag{59$'$}$$

we have, with (58), (63), and (63′) taken into account,

$$\begin{aligned} x(0) &= Q(0)Cx(0) + P(0)T(0)x(0) \\ &= Q(0)Cx(0) + P(0)z(0) - P(0)e(0) \\ &= Q(0)Cx(0) + P(0)T(0)x_0 - \tilde{e}(0) \\ &= [Q(0)C - I](x(0) - x_0) + x(0) - \tilde{e}(0), \end{aligned}$$

or equivalently

$$\tilde{e}(0) = [I - Q(0)C](x_0 - x(0)). \tag{64}$$

Thus by (56), we have

$$E\tilde{e}(0) = 0, \tag{65a}$$

and hence in view of (62′) and the assumption (3),

$$E\tilde{e}(t) = 0 \qquad \text{for} \quad t = 1, 2, \ldots . \tag{65b}$$

By virtue of (65), the covariance matrix of $\tilde{e}(t)$ becomes

$$\tilde{S}(t) \equiv \operatorname{cov}(\tilde{e}(t), \tilde{e}(t)) = E(\tilde{e}(t)\tilde{e}^T(t)), \qquad t = 1, 2, \ldots . \tag{66}$$

By virtue of (64), $\tilde{S}(0)$ is written as

$$\tilde{S}(0) = [I - Q(0)C]\Sigma(0)[I - Q(0)C]^T, \tag{67a}$$

where

$$\Sigma(0) \equiv E[(x_0 - x(0))(x_0 - x(0))^T], \tag{67b}$$

which is assumed to be positive semidefinite. $\tilde{S}(t)$ is parameterized as follows, by substituting (62′) into (66) and taking account of assumption (3), for $t = 1, 2, \ldots$:

$$\tilde{S}(t) = \Psi(Q(t), \tilde{S}(t-1)), \tag{68}$$

where

$$\begin{aligned}\Psi(&Q(t+1), \tilde{S}(t)) \\ &\equiv [I - Q(t+1)C][A\tilde{S}(t)A^T + AQ(t)\Phi Q^T(t)A^T + D\Xi D^T] \\ &\quad \times [I - Q(t+1)C]^T,\end{aligned} \tag{69}$$

since

$$\begin{aligned}E(\tilde{e}(t)\zeta^T(t)) &= [I - Q(t)C]AE(\tilde{e}(t-1)\zeta^T(t)) \\ &= [I - Q(t)C]A[I - Q(t-1)C]AE(\tilde{e}(t-2)\zeta^T(t)) \\ &\;\;\vdots \\ &= \prod_{\tau=1}^{t}([I - Q(\tau)C]A)P(0)E[(x_0 - x(0))\zeta^T(t)] \\ &= 0\end{aligned}$$

in view of (64) and (57′), and since similarly

$$E(\tilde{e}(t-1)\xi^T(t)) = 0.$$

Now we want to find $Q(t)$ minimizing the covariance matrix of estimation error $\tilde{e}(t)$. Define a subset of $n \times r$ matrices

$$\begin{aligned}\eta(\tilde{S}(t)) \equiv \{\, &Q(t+1) : Q(t+1)C[A\tilde{S}(t)A^T + AQ(t)\Phi Q^T(t)A^T \\ &\qquad + D\Xi D^T]C^T \\ &= [A\tilde{S}(t)A^T + AQ(t)\Phi Q^T(t)A^T + D\Xi D^T]C^T \,\}.\end{aligned} \tag{70}$$

Lemma 5 (Tse and Athans, 1970). *Let $\tilde{S}(t)$ be positive semidefinite. If $\tilde{Q}(t+1)$ belongs to $\eta(\tilde{S}(t))$, then*

$$\Psi(\tilde{Q}(t+1),\tilde{S}(t)) \leqslant \Psi(Q(t+1),\tilde{S}(t)) \qquad \textit{for all} \quad Q(t+1). \tag{71}$$

PROOF. Define

$$R(\tilde{S}(t)) \equiv C\big[A\tilde{S}(t)A^T + AQ(t)\Phi Q^T(t)A^T + D\Xi D^T\big]C^T. \tag{72}$$

Then $\eta(\tilde{S}(t))$ can be rewritten

$$\begin{aligned}\eta(\tilde{S}(t)) \equiv \big\{ Q(t+1) : {} & Q(t+1)R(\tilde{S}(t)) \\ &= \big[A\tilde{S}(t)A^T + AQ(t)\Phi Q^T(t)A^T + D\Xi D^T\big]C^T \big\}.\end{aligned}$$

We calculate

$$\begin{aligned}
&\Psi(\tilde{Q}(t+1),\tilde{S}(t)) + (\tilde{Q}(t+1) - Q(t+1))R(\tilde{S}(t)) \\
&\qquad \times(\tilde{Q}(t+1) - Q(t+1))^T \\
&= \Big\{\big[A\tilde{S}(t)A^T + AQ(t)\Phi Q^T(t)A^T + D\Xi D^T\big] \\
&\qquad -2\tilde{Q}(t+1)R(\tilde{S}(t))\tilde{Q}^T(t+1) + \tilde{Q}(t+1)R(\tilde{S}(t))\tilde{Q}^T(t+1)\Big\} \\
&\qquad +(\tilde{Q}(t+1) - Q(t+1))R(\tilde{S}(t))(\tilde{Q}(t+1) - Q(t+1))^T \\
&= \big[A\tilde{S}(t)A^T + AQ(t)\Phi Q^T(t)A^T + D\Xi D^T\big] \\
&\qquad - \tilde{Q}(t+1)R(\tilde{S}(t))\,Q^T(t+1) \\
&\qquad - Q(t+1)R(\tilde{S}(t))\,\tilde{Q}^T(t+1) + Q(t+1)R(\tilde{S}(t))\,Q^T(t+1) \\
&= \Psi(Q(t+1),\tilde{S}(t)).
\end{aligned} \tag{73}$$

Since $\tilde{S}(t)$ is positive semidefinite, (71) follows from (73). □

Corollary. *If $\tilde{Q}(t+1)$ and $\hat{Q}(t+1)$ belong to $\eta(\tilde{S}(t))$, then*

$$\Psi(\tilde{Q}(t+1),\tilde{S}(t)) = \Psi(\hat{Q}(t+1),\tilde{S}(t)). \tag{74}$$

PROOF. From (71) we have

$$\Psi(\tilde{Q}(t+1),\tilde{S}(t)) \leqslant \Psi(\hat{Q}(t+1),\tilde{S}(t))$$

and

$$\Psi(\hat{Q}(t+1),\tilde{S}(t)) \leqslant \Psi(\tilde{Q}(t+1),\tilde{S}(t)).$$ □

It is clear from (69) that if $\tilde{S}_1(t) \geqslant \tilde{S}_2(t)$, then

$$\Psi(Q(t+1),\tilde{S}_1(t)) \geqslant \Psi(Q(t+1),\tilde{S}_2(t)). \tag{75}$$

We adopt the optimality criterion that if covariance matrices of estimation errors are minimized in the sense of (71) for all t, the estimator is optimal.

Theorem 5 (Tse and Athans, 1970). a) *An optimum class of observer-estimators is specified by a sequence*

$$\{\tilde{Q}(t)\}_{t=1}^{\infty} \tag{76}$$

such that $\tilde{Q}(t) \in \eta(\tilde{S}^*(t-1))$ *for* $t = 1, 2, \ldots,$ *where* $\tilde{S}^*(t)$ *is determined uniquely by*

$$\tilde{S}^*(t) = \Delta(t-1) - \tilde{Q}(t)C\Delta(t-1) \qquad \text{for} \quad t = 1, 2, \ldots, \tag{77}$$

with $\tilde{S}^*(0) = \tilde{S}(0)$, *which is assumed to be positive semidefinite.* $\Delta(t)$ *is defined as*

$$\Delta(t) \equiv A\tilde{S}^*(t)A^T + A\tilde{Q}(t)\Phi\tilde{Q}^T(t)A^T + D\Xi D^T. \tag{78}$$

b) *If either* Φ *is positive definite or* $CD\Xi D^TC^T$ *is positive definite (or both), then a unique optimum class of observer-estimators exists and is specified by*

$$\tilde{Q}(t) = \Delta(t-1)C^T\big(C\Delta(t-1)C^T\big)^{-1}, \tag{79}$$

where $\Delta(t)$ *is that of* (78) *and* $\tilde{S}^*(t)$ *is determined recursively as*

$$\tilde{S}^*(t) = \Delta(t-1) - \Delta(t-1)C^T\big(C\Delta(t-1)C^T\big)^{-1}C\Delta(t-1). \tag{77'}$$

Proof. a) Given $Q(0)$ and $S(0)$, we see by (78)

$$\Delta(0) = A\tilde{S}(0)A^T + AQ(0)\Phi Q^T(0)A^T + D\Xi D^T, \tag{78'a}$$

where $\tilde{S}(0)$ is that given by (67a). Then by (68)

$$\tilde{S}^*(1) = \big[I - \tilde{Q}(1)C\big]\Delta(0)\big[I - \tilde{Q}(1)C\big]^T,$$

where $\tilde{Q}(1)$ belongs to $\eta(\tilde{S}(0))$, i.e., $\tilde{Q}(1)$ fulfills

$$\tilde{Q}(1)C\Delta(0)C^T = \Delta(0)C^T.$$

Hence $\tilde{S}^*(1)$ reduces to

$$\tilde{S}^*(1) = \Delta(0) - \tilde{Q}(1)C\Delta(0).$$

Next $\Delta(1)$ is obtained as

$$\Delta(1) = A\tilde{S}^*(1)A^T + A\tilde{Q}(1)\Phi\tilde{Q}^T(1)A^T + D\Xi D^T, \tag{78'b}$$

and for any matrix $\tilde{Q}(2)$ belonging to $\eta(\tilde{S}^*(1))$,

$$\tilde{Q}(2)C\Delta(1)C^T = \Delta(1)C^T.$$

Hence we have

$$\begin{aligned}\tilde{S}^*(2) &= \big[I - \tilde{Q}(2)C\big]\Delta(1)\big[I - \tilde{Q}(2)C\big]^T \\ &= \Delta(1) - \tilde{Q}(2)C\Delta(1).\end{aligned}$$

Proceeding this way, we get (77). By Lemma 5, $\tilde{S}^*(t)$ thus obtained at some t satisfies $\tilde{S}^*(t) \leqslant \tilde{S}(t)$. Then by (68), (75), and (71)

$$\begin{aligned}\tilde{S}(t+1) &= \Psi(Q(t+1), \tilde{S}(t)) \\ &\geqslant \Psi(Q(t+1), \tilde{S}^*(t)) \\ &\geqslant \Psi(\tilde{Q}(t+1), \tilde{S}^*(t)) \\ &= \tilde{S}^*(t+1).\end{aligned}$$

Thus $\tilde{S}^*(t) \leqslant \tilde{S}(t)$ for all t, implying optimality.

b) In general, there are more than one optimum class of observer-estimators which yield the same performance. If Φ is positive definite or $CD\Xi D^T C^T$ is positive definite (or both), then $C\Delta(t-1)C^T$ is positive definite. Hence from

$$\tilde{Q}(t)C\Delta(t-1)C^T = \Delta(t-1)C^T$$

we get the unique $\tilde{Q}(t)$ in (79). Substitution of (79) into (77) yields (77′). □

Note that (77′) has the same structure as covariance error matrix (49) of Kalman filter. Note also that in Theorem 5 the noise covariance matrix Φ and the initial state distribution covariance matrix $\tilde{S}(0)$ are assumed to be positive definite and positive semidefinite, respectively. See Yoshikawa (1975) for the minimal-order optimal filters for linear discrete-time systems without these assumptions.

4.4. Minimal-Order Observer-Estimators: Computation

In order to implement the rules of Theorem 5 in the practice of estimation, suppose the first r variables of state vector x are observed (with or without observation error), i.e., matrix C in (2) is specified as (cf. the end of Section 3.1.)

$$C = [I_r, 0], \tag{80}$$

where I_r stands for the identity matrix of order r. Then $\Delta(t)C^T$ and $C\Delta(t)C^T$ are reduced to

$$\Delta(t)C^T = \Delta_\dagger^T(t) \equiv A\tilde{S}^*(t)A_\dagger^T + A\tilde{Q}(t)\Phi\tilde{Q}^T(t)A_\dagger^T + D\Xi D_\dagger^T, \tag{81}$$

and

$$\begin{aligned}C\Delta(t)C^T &= \Delta_r(t) \\ &\equiv A_\dagger\tilde{S}^*(t)A_\dagger^T + A_\dagger\tilde{Q}(t)\Phi\tilde{Q}^T(t)A_\dagger^T + D_\dagger\Xi D_\dagger^T,\end{aligned} \tag{82}$$

where $A_\dagger$ and $D_\dagger$ denote $r \times n$ submatrices of A and D, respectively, formed by their first r rows. Provided $\Delta_r(t)$ is positive definite, we get from

(79) and (77′)

$$\tilde{Q}(t+1) = \Delta_{\dagger}^T(t)\Delta_r(t)^{-1} = \begin{pmatrix} I_r \\ \tilde{G}(t) \end{pmatrix} \tag{83}$$

and

$$\tilde{S}^*(t+1) = \Delta(t) - \Delta_{\dagger}^T(t)\Delta_r(t)^{-1}\Delta_{\dagger}(t) = \begin{pmatrix} 0 & 0 \\ 0 & \sigma_s(t) \end{pmatrix}, \tag{84}$$

where (for singular $\Delta_r(t)$, its generalized inverse replaces $\Delta_r(t)^{-1}$)

$$\tilde{G}(t) \equiv \Delta_{rs}^T(t)\Delta_r(t)^{-1}, \tag{85}$$

$$\sigma_s(t) \equiv \Delta_s(t) - \Delta_{rs}^T(t)\Delta_r(t)^{-1}\Delta_{rs}(t). \tag{86}$$

$\Delta_{rs}(t)$ is the $r \times s$ submatrix of $\Delta_{\dagger}(t)$ formed by its last s columns, and $\Delta_s(t)$ is the $s \times s$ submatrix of $\Delta(t)$ formed by its last s rows and columns.

Putting

$$P(t) = \begin{pmatrix} 0 \\ I_s \end{pmatrix}, \tag{87}$$

we have from (58), in view of (80) and (83),

$$T(t) = \left[-\tilde{G}(t), I_s \right]. \tag{88}$$

Therefore, our order s observer-estimator (60) can be written, with (61) taken into consideration,

$$\begin{aligned} z(t) &= (A_{22} - \tilde{G}(t)A_{12})z(t-1) + (B_{22} - \tilde{G}(t)B_{11})v(t) \\ &\quad + \left[A_{21} - \tilde{G}(t)A_{11} + (A_{22} - \tilde{G}(t)A_{12})\tilde{G}(t-1) \right] y(t-1), \end{aligned} \tag{89}$$

where we use the following partitioned matrices:

$$A = \begin{pmatrix} A_{11} & A_{12} \\ A_{21} & A_{22} \end{pmatrix} \begin{matrix} \}r \\ \}s \end{matrix} \quad \text{(columns: } r, s\text{)} \qquad B = \begin{pmatrix} B_{11} \\ B_{22} \end{pmatrix} \begin{matrix} \}r \\ \}s \end{matrix} \tag{90}$$

Under the specification (80), $\hat{x}(t)$ in (59) will be expressed as

$$\hat{x}(t) = \begin{pmatrix} y(t) \\ \hat{w}(t) \end{pmatrix} \tag{91}$$

where $\hat{w}(t)$ denotes the estimator for unobserved vector $w(t)$ in a state vector $x(t)$. Thus, in view of (88), we get

$$\hat{w}(t) = z(t) + \tilde{G}(t)y(t) \tag{92}$$

from (59). Substitution of (92) into (89) yields

$$\begin{aligned} \hat{w}(t) &= (A_{22} - \tilde{G}(t)A_{12})\hat{w}(t-1) + (A_{21} - \tilde{G}(t)A_{11})y(t-1) \\ &\quad + \tilde{G}(t)y(t) + (B_{22} - \tilde{G}(t)B_{11})v(t). \end{aligned} \tag{93}$$

To initiate the computation, assume

$$\tilde{G}(0) = G(0) = 0, \tag{94}$$

which implies, in view of (63) and (59′), that

$$e(0) = [0, I_s](x_0 - x(0)), \tag{94′}$$

and in view of (67a), that

$$\tilde{S}^*(0) = \tilde{S}(0) = \begin{pmatrix} 0 & 0 \\ 0 & S(0) \end{pmatrix}, \tag{84′a}$$

where

$$S(0) \equiv E\big(e(0)e^T(0)\big). \tag{94″}$$

Then by (78′a)

$$\Delta(0) \equiv \begin{pmatrix} \Delta_r(0) & \Delta_{rs}(0) \\ \Delta_{rs}^T(0) & \Delta_s(0) \end{pmatrix}$$

$$= \begin{bmatrix} A_{12}S(0)A_{12}^T + A_{11}\Phi A_{11}^T & A_{12}S(0)A_{22}^T + A_{11}\Phi A_{21}^T \\ A_{22}S(0)A_{12}^T + A_{21}\Phi A_{11}^T & A_{22}S(0)A_{22}^T + A_{21}\Phi A_{21}^T \end{bmatrix} + D\Xi D^T \tag{78″a}$$

Thus by (86) we calculate $\sigma_s(0)$ and then by (84)

$$\tilde{S}^*(1) = \begin{pmatrix} 0 & 0 \\ 0 & \sigma_s(0) \end{pmatrix}, \tag{84′b}$$

and, by (83) and (85), we have

$$\tilde{Q}(1) = \begin{pmatrix} I_r \\ \Delta_{rs}^T(0)\Delta_r(0)^{-1} \end{pmatrix}. \tag{79′a}$$

(If $\Delta_r(0)$ is not invertible, its generalized inverse will replace $\Delta_r(0)^{-1}$.)

Now we use (78′b) to compute $\Delta(1)$, and then by (85)

$$\tilde{G}(1) = \Delta_{rs}^T(1)\Delta_r(1)^{-1}. \tag{85′a}$$

The initial value of $\hat{w}(0)$ will be picked as

$$\hat{w}(0) = [0, I_s]x_0 \tag{93′a}$$

in view of (59′) and (94). Thus we get our observer-estimator in the first period

$$\begin{aligned} \hat{w}(1) = {} & (A_{22} - \tilde{G}(1)A_{12})[0, I_s]x_0 + (A_{21} - \tilde{G}(1)A_{11})y(0) \\ & + \tilde{G}(1)y(1) + (B_{22} - \tilde{G}(1)B_{11})v(1), \end{aligned} \tag{93′b}$$

to which the above $\tilde{G}(1)$ is applied.

Next, for $t = 2$, $S^*(2)$ is obtained as

$$\tilde{S}^*(2) = \begin{pmatrix} 0 & 0 \\ 0 & \sigma_s(1) \end{pmatrix}, \tag{84′c}$$

where

$$\sigma_s(1) \equiv \Delta_s(1) - \Delta_{rs}^T(1)\Delta_r(1)^{-1}\Delta_{rs}(1). \tag{86′a}$$

$\tilde{Q}(2)$ is already known in view of (83) and (85′a). Thus $\Delta(2)$ is calculated by applying (78), and hence $\tilde{G}(2)$ can be obtained by (85). So the observer-estimator $\hat{w}(2)$ will be computed as

$$\begin{aligned}\hat{w}(2) = \left(A_{22} - \tilde{G}(2)A_{12}\right)\hat{w}(1) + \left(A_{21} - \tilde{G}(2)A_{11}\right)y(1) \\ + \tilde{G}(2)y(2) + \left(B_{22} - \tilde{G}(2)B_{11}\right)v(2).\end{aligned} \tag{93′c}$$

For subsequent $t = 3, 4, \ldots,$ we proceed one by one in this manner.

Application 2. Our minimal-order observer-estimator will be found useful for such a generalized distributed-lag system as

$$x(t) = \sum_{i=1}^{k} A_i x(t-i) + \sum_{j=0}^{h} B_j v(t-j) + \xi(t) \tag{95}$$

with output system

$$y(t) = Cx(t) + \zeta(t) \tag{96}$$

where $x(t)$, $v(t)$, and $y(t)$ are the state n vector, control m vector, and output r vector, respectively, in period t; $\xi(t)$, $\zeta(t)$ are stochastic terms obeying assumptions (3); and A_i, B_j, C are constant matrices of appropriate dimensions with $rk(C) = r < n$. Equation (95) can be rewritten

$$\tilde{x}(t) = \tilde{A}\tilde{x}(t-1) + \tilde{B}v(t) + \tilde{D}\xi(t), \tag{95′}$$

where

$$\tilde{x}(t) \equiv \begin{bmatrix} v(t-h+1) \\ \vdots \\ v(t-1) \\ v(t) \\ x(t-k+1) \\ \vdots \\ x(t-1) \\ x(t) \end{bmatrix}, \quad \tilde{A} \equiv \begin{bmatrix} 0 & & I_m & 0 & 0 & \cdots & \cdots & 0 \\ \vdots & \ddots & & \ddots & \vdots & & & \vdots \\ \vdots & & 0 & I_m & 0 & \cdots & \cdots & 0 \\ 0 & \cdots & 0 & 0 & 0 & \cdots & \cdots & 0 \\ 0 & \cdots & \cdots & 0 & 0 & & I_n & 0 \\ \vdots & & & & & \ddots & & \ddots \\ 0 & \cdots & \cdots & 0 & 0 & \cdots & 0 & I_n \\ B_h & \cdots & \cdots & B_1 & A_k & \cdots & A_2 & A_1 \end{bmatrix},$$

$$\tilde{B} \equiv \begin{bmatrix} 0 \\ \vdots \\ 0 \\ I_m \\ 0 \\ \vdots \\ 0 \\ B_0 \end{bmatrix}, \qquad \tilde{D} \equiv \begin{bmatrix} 0 \\ \vdots \\ \vdots \\ \vdots \\ \vdots \\ \vdots \\ 0 \\ I_n \end{bmatrix}.$$

Accordingly, (96) is expressed as

$$\tilde{y}(t) = \tilde{C}\tilde{x}(t) + F\xi(t), \tag{96'}$$

where

$$\tilde{y}(t) \equiv \begin{bmatrix} v(t-h+1) \\ \vdots \\ v(t-1) \\ v(t) \\ x(t-k+1) \\ \vdots \\ x(t-1) \\ y(t) \end{bmatrix},$$

$$\tilde{C} \equiv \begin{bmatrix} I_m & & (h) & & & & \\ & \ddots & & & & & \\ & & I_m & & & & \\ & & & I_m & & (k-1) & \\ & & & & I_n & & \\ & & & & & \ddots & \\ & & & & & & I_n & \\ & & & & & & & C \end{bmatrix}, \qquad F \equiv \begin{bmatrix} 0 \\ \vdots \\ \vdots \\ \vdots \\ \vdots \\ 0 \\ I_r \end{bmatrix}.$$

Note that the dimensions of $\tilde{x}(t)$ and $\tilde{y}(t)$ are $\tilde{n} \equiv hm + kn$ and $\tilde{r} \equiv hm + kn - n + r$, respectively. Clearly, $rk(\tilde{C}) = \tilde{r} < \tilde{n}$. Thus the transformed state-space representation (95′), (96′) is of the same structure as system (1), (2). If C is given by

$$C = [I_r, 0], \tag{80}$$

then $\tilde{C}$ will become

$$\tilde{C} = [I_{\tilde{r}}, 0], \tag{80'}$$

and we can follow the previous procedure for obtaining the observer-estimator for the vector of unobserved $n - r$ variables in $x(t)$.

We shall compute the minimal-order observer-estimator for x_2 in the simple system composed of the one-period lag state equation

$$\begin{pmatrix} x_1(t) \\ x_2(t) \end{pmatrix} = \begin{pmatrix} a_{11} & a_{12} \\ a_{21} & a_{22} \end{pmatrix} \begin{pmatrix} x_1(t-1) \\ x_2(t-1) \end{pmatrix} + \begin{pmatrix} b_{11}^0 \\ b_{22}^0 \end{pmatrix} v(t) + \begin{pmatrix} b_{11}^1 \\ b_{22}^1 \end{pmatrix} v(t-1) + \begin{pmatrix} \xi_1(t) \\ \xi_2(t) \end{pmatrix} \tag{13*}$$

and the output equation

$$y(t)=C\begin{pmatrix}x_1(t)\\ x_2(t)\end{pmatrix}+\zeta(t),\qquad \text{with}\quad C=[1,0]. \tag{14*}$$

This is the special case of system (95), (96) where we set state-vector dimension $n=2$, control-vector dimension $m=1$, and output-vector dimension $r=1$, together with $k=1$ and $h=1$. Coefficients a_{ij}, b_{ii}^0, b_{ii}^1 $(i,j,=1,2)$ and variables $x_i(\tau)$, $v(\tau)$ $(\tau=t,t-1)$, $\xi_i(\tau)$ $(i=1,2)$, $y(t)$, $\zeta(t)$ are all scalars. The state-space form of the present system is shown below in view of (95′) and (96′):

$$\tilde{x}(t)=\tilde{A}\tilde{x}(t-1)+\tilde{B}v(t)+\tilde{D}\xi(t), \tag{15*}$$

$$\tilde{y}(t)=\tilde{C}\tilde{x}(t)+F\zeta(t), \tag{16*}$$

where

$$\tilde{x}(t)\equiv\begin{bmatrix}v(t)\\ x_1(t)\\ x_2(t)\end{bmatrix},\quad \tilde{A}\equiv\begin{bmatrix}0&0&0\\ b_{11}^1&a_{11}&a_{12}\\ b_{22}^1&a_{21}&a_{22}\end{bmatrix},\quad \tilde{B}\equiv\begin{bmatrix}1\\ b_{11}^0\\ b_{22}^0\end{bmatrix},\quad \tilde{D}\equiv\begin{bmatrix}0&0\\ 1&0\\ 0&1\end{bmatrix}, \tag{17*a}$$

$$\tilde{y}(t)\equiv\begin{pmatrix}v(t)\\ y(t)\end{pmatrix},\quad \tilde{C}\equiv\begin{pmatrix}1&0&0\\ 0&1&0\end{pmatrix},\quad F\equiv\begin{pmatrix}0\\ 1\end{pmatrix},\quad \xi(t)\equiv\begin{pmatrix}\xi_1(t)\\ \xi_2(t)\end{pmatrix}. \tag{17*b}$$

Note that the dimension $\tilde{n}$ of the new state vector $\tilde{x}(t)$ is equal to three, and that the dimension $\tilde{r}$ of the new output vector $\tilde{y}(t)$ is equal to two. So $\tilde{s}\equiv\tilde{n}-\tilde{r}=1$. Hereafter, the notations r and s will replace $\tilde{r}$ and $\tilde{s}$, respectively, for the sake of brevity. We assume that the variances of stochastic disturbances $\zeta(t)$ and $\xi(t)$ are given by

$$\operatorname{var}(\zeta(t))=\Phi>0,\qquad \operatorname{var}(\xi(t))=\Xi=\begin{pmatrix}\Xi_1&0\\ 0&\Xi_2\end{pmatrix},\qquad (\Xi_1,\Xi_2>0). \tag{18*}$$

We want to obtain the observer-estimator for inaccessible variable $x_2(t)$, which is formally given by $\hat{w}(t)$ in (93), by calculating $\hat{w}(0)$, $\hat{w}(1)$, $\hat{w}(2)$, and so on. $\hat{w}(0)$ for the present example is easily seen to be

$$\hat{w}(0)=x_2^0 \tag{19*a}$$

by virtue of (93′a), where x_2^0 is a given initial value of x_2. Noticing that matrix $\tilde{A}$ and $\tilde{B}$ are partitioned as

$$\tilde{A}=\left[\begin{array}{cc:c}0&0&0\\ b_{11}^1&a_{11}&a_{12}\\ \hdashline b_{22}^1&a_{21}&a_{22}\end{array}\right]\equiv\left(\begin{array}{c:c}A_{11}&A_{12}\\ \hdashline A_{21}&A_{22}\end{array}\right),\qquad \tilde{B}=\left[\begin{array}{c}1\\ b_{11}^0\\ \hdashline b_{22}^0\end{array}\right]\equiv\left(\begin{array}{c}B_{11}\\ \hdashline B_{22}\end{array}\right), \tag{20*}$$

we can derive $\hat{w}(1)$ by (93′b) as

$$\begin{aligned}\hat{w}(1)=&\,(a_{22}-a_{12}q(1))\hat{w}(0)+(b_{22}^1-b_{11}^1q(1))v(0)\\ &+(a_{21}-a_{11}q(1))y(0)+q(1)y(1)+(b_{22}^0-b_{11}^0q(1))v(1),\end{aligned} \tag{19*b}$$

where

$$q(1) \equiv \alpha_{21}(1)/\alpha_{11}(1), \tag{21*a}$$

$$\alpha_{11}(1) \equiv \sigma_s(0)(a_{12})^2 + \Phi\left[(a_{11} + a_{12}q(0))^2 + (b_{11}^1)^2\right] + \Xi_1, \tag{22*a}$$

$$\alpha_{21}(1) \equiv \sigma_s(0)a_{12}a_{22} + \Phi\left[(a_{11} + a_{12}q(0))(a_{21} + a_{22}q(0)) + b_{11}^1 b_{22}^1\right], \tag{22*b}$$

$$q(0) \equiv \alpha_{21}(0)/\alpha_{11}(0), \tag{23*}$$

$$\sigma_s(0) \equiv \alpha_{22}(0) - q(0)\alpha_{21}(0), \tag{24*}$$

$$\alpha_{11}(0) \equiv S(0)(a_{12})^2 + \Phi\left[(a_{11})^2 + (b_{11}^1)^2\right] + \Xi_1, \tag{25*a}$$

$$\alpha_{21}(0) \equiv S(0)a_{12}a_{22} + \Phi\left[a_{11}a_{21} + b_{11}^1 b_{22}^1\right], \tag{25*b}$$

$$\alpha_{22}(0) \equiv S(0)(a_{22})^2 + \Phi\left[(a_{21})^2 + (b_{22}^1)^2\right] + \Xi_2, \tag{25*c}$$

$$S(0) \equiv E\left(x_2^0 - x_2(0)\right)^2, \qquad \text{(in view of (94$'$) and (94$''$))}. \tag{26*}$$

The derivational procedure for (19*b) is as follows. By (78″a) and (20*), we get

$$\Delta_r(0) = A_{12}S(0)A_{12}^T + A_{11}\Phi A_{11}^T + \begin{pmatrix} 0 & 0 \\ 0 & \Xi_1 \end{pmatrix} = \begin{pmatrix} 0 & 0 \\ 0 & \alpha_{11}(0) \end{pmatrix}, \tag{27*a}$$

$$\Delta_{rs}^T(0) = A_{22}S(0)A_{12}^T + A_{21}\Phi A_{11}^T = (0, \alpha_{21}(0)), \tag{27*b}$$

$$\Delta_s(0) = A_{22}S(0)A_{22}^T + A_{21}\Phi A_{21}^T + \Xi_2 = \alpha_{22}(0). \tag{27*c}$$

Since $\Delta_r(0)$ in (27*a) is not invertible, we obtain its generalized inverse as

$$\Delta_r(0)^+ = \begin{pmatrix} 0 & 0 \\ 0 & \alpha_{11}(0)^{-1} \end{pmatrix} \tag{28*}$$

by applying the corollary to Theorem 22 of Murata (1977, p. 215). Thus we have

$$\Delta_{rs}^T(0)\Delta_r(0)^+ = (0, q(0)). \tag{29*a}$$

Hence $\tilde{Q}(1)$ in (79′a) becomes

$$\tilde{Q}(1) = \begin{bmatrix} 1 & 0 \\ 0 & 1 \\ 0 & q(0) \end{bmatrix}, \tag{30*}$$

and $\sigma_s(0)$ in (86) with $t = 0$ becomes

$$\sigma_s(0) = \Delta_s(0) - \Delta_{rs}^T(0)\Delta_r(0)^+\Delta_{rs}(0) \tag{31*a}$$

which is equal to (24*). Taking (84′b) into account, therefore, we get by (78′b)

$$\Delta(1) = \tilde{A}\tilde{S}^*(1)\tilde{A}^T + \tilde{A}\tilde{Q}(1)\Phi\tilde{Q}^T(1)\tilde{A}^T + \tilde{D}\Xi\tilde{D}^T$$
$$= \left[\begin{array}{cc:c} 0 & 0 & 0 \\ 0 & \alpha_{11}(1) & \alpha_{21}(1) \\ \hdashline 0 & \alpha_{21}(1) & \alpha_{22}(1) \end{array}\right] \equiv \left(\begin{array}{c:c} \Delta_r(1) & \Delta_{rs}(1) \\ \hdashline \Delta_{rs}^T(1) & \Delta_s(1) \end{array}\right) \tag{32*}$$

where $\alpha_{11}(1)$ and $\alpha_{21}(1)$ are given by (22*a) and (22*b) above, and

$$\alpha_{22}(1) \equiv \sigma_s(0)(a_{22})^2 + \Phi\Big[\big(a_{21} + a_{22}q(0)\big)^2 + \big(b_{22}^1\big)^2\Big] + \Xi_2. \tag{22*c}$$

Thus $\tilde{G}(1)$ in (85′a) becomes

$$\tilde{G}(1) = \Delta_{rs}^T(1)\Delta_r(1)^+ = (0, q(1)). \tag{29*b}$$

Substituting (29*b) and (20*) into (93′b) yields (19*b).

In a similar manner, we get from (93′c)

$$\hat{w}(2) = (a_{22} - a_{12}q(2))\hat{w}(1) + \big(b_{22}^1 - b_{11}^1 q(2)\big)v(1)$$
$$+ (a_{21} - a_{11}q(2))y(1) + q(2)y(2) + \big(b_{22}^0 - b_{11}^0 q(2)\big)v(2), \tag{19*c}$$

where

$$q(2) \equiv \alpha_{21}(2)/\alpha_{11}(2), \tag{21*b}$$

$$\alpha_{11}(2) \equiv \sigma_s(1)(a_{12})^2 + \Phi\Big[\big(a_{11} + a_{12}q(1)\big)^2 + \big(b_{11}^1\big)^2\Big] + \Xi_1, \tag{33*a}$$

$$\alpha_{21}(2) \equiv \sigma_s(1)a_{12}a_{22} + \Phi\big[\big(a_{11} + a_{12}q(1)\big)\big(a_{21} + a_{22}q(1)\big) + b_{11}^1 b_{22}^1\big], \tag{33*b}$$

$$\sigma_s(1) \equiv \alpha_{22}(1) - q(1)\alpha_{21}(1). \tag{31*b}$$

Thus we can deduce the following iterative form of the observer-estimator for variable $x_2(t)$ in the present problem:

$$\hat{w}(t) = (a_{22} - a_{12}q(t))\hat{w}(t-1) + \big(b_{22}^1 - b_{11}^1 q(t)\big)v(t-1)$$
$$+ (a_{21} - a_{11}q(t))y(t-1) + q(t)y(t) + \big(b_{22}^0 - b_{11}^0 q(t)\big)v(t) \tag{34*}$$

for $t = 1, 2, \ldots,$ where

$$q(t) \equiv \alpha_{21}(t)/\alpha_{11}(t), \tag{35*}$$

$$\alpha_{11}(t) \equiv \sigma_s(t-1)(a_{12})^2 + \Phi\Big[\big(a_{11} + a_{12}q(t-1)\big)^2 + \big(b_{11}^1\big)^2\Big] + \Xi_1, \tag{36*a}$$

$$\alpha_{21}(t) \equiv \sigma_s(t-1)a_{12}a_{22}$$
$$+ \Phi\big[\big(a_{11} + a_{12}q(t-1)\big)\big(a_{21} + a_{22}q(t-1)\big) + b_{11}^1 b_{22}^1\big], \tag{36*b}$$

$$\sigma_s(t) \equiv \alpha_{22}(t) - q(t)\alpha_{21}(t), \tag{37*}$$

$$\alpha_{22}(t) \equiv \sigma_s(t-1)(a_{22})^2 + \Phi\Big[\big(a_{21} + a_{22}q(t-1)\big)^2 + \big(b_{22}^1\big)^2\Big] + \Xi_2. \tag{36*c}$$

4.5. Economic Applications of Kalman Filtering Methods

Perhaps the earliest application of the Kalman filter to economics was made by Taylor (1970), who introduced an optimal production rule in relation to inventory and production costs. In his example, the long-run demands $x(t)$ for the firm's inventoried goods are unobserved, while observed demands $y(t)$ are perturbed from the long-run demand levels $Cx(t)$ by the transitory demand component $\zeta(t)$, and hence our equation (2) holds. The long-run demands are assumed to be stochastic such that

$$x(t) = Ax(t-1) + D\xi(t). \tag{97}$$

Assuming (3) as for the stochastic components $\xi(t)$ and $\zeta(t)$, therefore, we can directly apply the Kalman filtering algorithm (48) to estimating $x(t)$, i.e.,

$$\hat{x}(t\mid t) = \left[I - N(t)C\right]A\hat{x}(t-1\mid t-1) + N(t)y(t), \tag{98}$$

where $N(t)$ is that of (38), given some initial values. Taylor (1970) also gave a stochastic optimal control rule, as is seen in Section 5.1 in conjunction with estimate $\hat{x}(t\mid t)$. He thus demonstrated the separation principle of estimation and control in stochastic circumstances (cf. Section 5.2), in parallel with the separation of observer and controller in deterministic cases (Section 3.5). Here we refer to Pagan (1975) who used the Kalman filter formula (98) in extracting components from time series.

Kalman filtering methods can be viewed as supplementary to econometric methods in the following two respects. First, as Vishwakarma (1974) suggested, in order to derive the future values of the exogenous variables needed to predict macroeconomic activities from an econometric model, the Kalman filter provides a convenient tool. Second, as Athans (1974) proposed, from a set of newly obtained values of all variables in an estimated econometric model, we can compute the updated estimates of parameters by using the Kalman filter algorithm. Since the former is clearly a straightforward application of the Kalman filter, we shall dwell only on the latter.

Application 3 (Updating Estimates of Parameters in an Econometric Model). We consider a general distributed-lag model in reduced form:

$$y(t) = \sum_{i=1}^{r} A_i y(t-i) + \sum_{j=1}^{s} B_j v(t-j+1) + b_0 + u(t), \tag{38*}$$

where

- y column n vector of endogenous (output) variables,
- v column m vector of exogenous (input) variables,
- b_0 constant column n vector,

u column n vector of stochastic errors, each component having zero mean and a finite variance,
A_i constant $n \times n$ matrix for $i = 1, 2, \ldots, r$,
B_j constant $n \times m$ matrix for $j = 1, 2, \ldots, s$.

Suppose that the parameter estimates of A_i, B_j, and b_0 are known by an econometric method but that these values are expected to vary as a set of new data becomes available for the variables in (38*), with the structure of the system remaining unchanged. Let $y_k(t)$ and $u_k(t)$ be the kth components of vectors $y(t)$ and $u(t)$, respectively, let a_i^k and b_j^k denote the kth row vectors of matrices A_i and B_j, respectively, and let b_0^k be the kth element of b_0. Then, equation (38*) is rewritten as

$$y_k(t) = \sum_{i=1}^{r} a_i^k y(t-i) + \sum_{j=1}^{s} b_j^k v(t-j+1) + b_0^k + u_k(t), \qquad k = 1, \ldots, n$$

or, equivalently,

$$y_k(t) = Cx_k(t) + u_k(t), \qquad k = 1, \ldots, n, \tag{39*}$$

where

$$x_k(t) \equiv (a_1^k, \ldots, a_r^k, b_1^k, \ldots, b_s^k, b_0^k), \tag{40*}$$

$$C \equiv \big(y^T(t-1), \ldots, y^T(t-r), v^T(t), \ldots, v^T(t-s+1), 1\big). \tag{41*}$$

Note that given a set of new data for C in (41*) and for $y_k(t)$, new parameters to be estimated are computed as the Kalman filter for $x_k(t)$ satisfying the trivial difference equation

$$x_k(t) = x_k(t-1) \qquad \text{for all } t. \tag{42*}$$

Thus, in the present problem, the Kalman filter (48) reduces to

$$\hat{x}_k(t \mid t) = \big[I - N_k(t)C\big]\hat{x}_k(t-1 \mid t-1) + N_k(t)y_k(t), \tag{43*}$$

where

$$N_k(t) \equiv \left(\frac{1}{CS_k(t)C^T + \Phi_k}\right) S_k(t)C^T \qquad (\Phi_k \equiv \text{var}(u_k(t)))$$

$$S_k(t) \equiv \big[S_k(t-1)^{-1} + C^T\Phi_k^{-1}C\big]^{-1}$$

with initial values:

$$\hat{x}_k(0 \mid 0) = \text{preliminary estimate } x_k(0) \text{ of } \big(a_1^k, \ldots, a_r^k, b_1^k, \ldots, b_s^k, b_0^k\big),$$

$$S_k(0) = \text{prior covariance matrix cov}(x_k(0), x_k(0)).$$

Using the recursive formula (43*), we can converge to a unique vector as the updated estimate of $x_k(t)$ in (40*). Finally, we notice a similar application of the Kalman filter to estimating Leontief matrix in an interindustry demand scheme (see Chow (1975, pp. 192–193)). For a further development along the lines of Athans (1974), refer to Lazaridis (1980).

Application 4 (Optimal Short-Run Monetary Policy). LeRoy and Waud (1977) convince us "that the Kalman filter is an indispensable analytical tool for the solution of short-run monetary control problems," because the current demand for money is an unobservable but essential variable in their policy model. Though their model does not contain explicitly any lag in time, we shall add on a one-period lagged variable to the model, to make clearer their intention of showing that the central bank's reserves policy is optimally determined in an intimate connection with the Kalman filter for money demand. Thus our model consists of the following two behavioral equations:

$$m_t = a_0 + a_1 i_t + a_2 m_{t-1} + u_{1t} \tag{44*}$$

$$r_t = b_0 + b_1 m_t + b_2 i_t + u_{2t}, \tag{45*}$$

where m_t, i_t, and r_t are the money stock, interest rate, and reserves in period t, m_{t-1} is the money stock lagged by one period (which is the additional variable mentioned above), and u_{1t} and u_{2t} are normally distributed and mutually uncorrelated disturbances with zero means and finite variances:

$$\operatorname{var}(u_{it}) = \sigma_i^2, \qquad i = 1, 2. \tag{46*}$$

We assume that the monetary authority has perfect knowledge of the parameters a_j and b_j $(j = 0, 1, 2)$ and the variances in (46*). Equation (44*) is rationalized as follows. Based on the money stock existing at the beginning of this period, "money holders communicate their desired money balances to banks in the form of loan applications contingent on the interest rate." Equation (45*) represents the banking system's behavior such that its "desired excess reserves depend exclusively on the interest rate and that its ability to meet the demand for money depends on the amount of reserves it has."

Suppose that the objective of the present monetary policy is to minimize the expected squared deviation of the money stock around some target level m^*. Assume that the current realizations of u_{1t} and u_{2t} have already occurred but that the central bank cannot observe them individually. Then, the desired demand for money is derived as a nonrandom value (which is not yet observed by the central bank) by (44*) for a given interest rate, and the banks in turn calculate their desired reserves and submit their demands for reserves contingent on the interest rate to the central bank (as (45*) indicates) which then, acting as the Walrasian auctioneer, can determine the quantity of reserves desired at the interest rate. This entire process is summarized mathematically by the following reduced form equation for r_t:

$$r_t = b_0 + a_0 b_1 + a_2 b_1 m_{t-1} + (b_2 + a_1 b_1) i_t + (u_{2t} + b_1 u_{1t}). \tag{47*}$$

In conjunction with its acquired knowledge of r_t and i_t, the central bank solves (47*) for the quantity $u_{2t} + b_1 u_{1t}$. In order to determine an optimal monetary policy, the central bank must use this quantity to estimate u_{1t}, i.e., to construct the estimate of u_{1t} conditional on $u_{2t} + b_1 u_{1t}$, which is given by[1]

$$E(u_{1t} \mid u_{2t} + b_1 u_{1t}) = \frac{b_1 \sigma_1^2 (u_{2t} + b_1 u_{1t})}{b_1^2 \sigma_1^2 + \sigma_2^2}. \tag{48*}$$

Using (48*), the central bank will obtain the conditional estimate of the unobservable value of m_t as

$$\hat{m}_t = a_0 + a_1 i_t + a_2 m_{t-1} + E(u_{1t} \mid u_{2t} + b_1 u_{1t}). \tag{49*}$$

We shall show that $\hat{m}_t$ in (49*) is nothing but the Kalman filter for m_t when m_{t-1} is regarded as a known initial value. The state equation of the present model is

$$\begin{pmatrix} m_t \\ r_t \end{pmatrix} = \begin{pmatrix} a_2 & 0 \\ a_2 b_1 & 0 \end{pmatrix} \begin{pmatrix} m_{t-1} \\ r_{t-1} \end{pmatrix} + \begin{pmatrix} a_1 \\ a_1 b_1 + b_2 \end{pmatrix} i_t + \begin{pmatrix} a_0 \\ a_0 b_1 + b_0 \end{pmatrix} + \begin{pmatrix} 1 & 0 \\ b_1 & 1 \end{pmatrix} \begin{pmatrix} u_{1t} \\ u_{2t} \end{pmatrix} \tag{50*}$$

which is the combination of the reduced form equations (44*) and (47*), and the output equation is

$$r_t = (0, 1) \begin{pmatrix} m_t \\ r_t \end{pmatrix}. \tag{51*}$$

Denoting $y(t) \equiv r_t$, $v(t) \equiv i_t$, $C \equiv (0, 1)$, and

$$x(t) \equiv \begin{pmatrix} m_t \\ r_t \end{pmatrix}, \quad A \equiv \begin{pmatrix} a_2 & 0 \\ a_2 b_1 & 0 \end{pmatrix}, \quad B \equiv \begin{pmatrix} a_1 \\ a_1 b_1 + b_2 \end{pmatrix},$$

$$k \equiv \begin{pmatrix} a_0 \\ a_0 b_1 + b_0 \end{pmatrix}, \quad D \equiv \begin{pmatrix} 1 & 0 \\ b_1 & 1 \end{pmatrix}, \quad \xi(t) \equiv \begin{pmatrix} u_{1t} \\ u_{2t} \end{pmatrix},$$

and setting the initial value at

$$\hat{x}(t-1 \mid t-1) = x(t-1),$$

we apply the Kalman filter formula (48) to our problem to yield

$$\hat{x}(t \mid t) = [I - N(t)C](Ax(t-1) + Bv(t) + k) + N(t)y(t), \tag{52*}$$

where

$$N(t) \equiv S(t)C^T(CS(t)C^T)^{-1} \tag{53*}$$

$$S(t) \equiv D \cdot \operatorname{cov}(\xi(t), \xi(t)) \cdot D^T = \begin{pmatrix} \sigma_1^2 & b_1 \sigma_1^2 \\ b_1 \sigma_1^2 & b_1^2 \sigma_1^2 + \sigma_2^2 \end{pmatrix}. \tag{54*}$$

[1] Let x_1 and x_2 be normally distributed random variables with $E(x_i) = \mu_i$, $\operatorname{var}(x_i) = \sigma_{ii}$ for $i = 1, 2$, and $\operatorname{cov}(x_1, x_2) = \sigma_{12}$. Then the distribution of x_1 conditional on x_2 has mean and variance as follows:

$$E(x_1 \mid x_2) = \mu_1 + \frac{\sigma_{12}}{\sigma_{22}}(x_2 - \mu_2), \quad \operatorname{var}(x_1 \mid x_2) = \sigma_{11} - \frac{\sigma_{12}^2}{\sigma_{22}}.$$

Hence $N^T(t) = (\sigma_n, 1)$ with

$$\sigma_n \equiv \frac{b_1\sigma_1^2}{b_1^2\sigma_1^2 + \sigma_2^2}. \tag{55*}$$

Thus (52*) implies that the estimate of r_t is indeed the observed r_t itself and that the Kalman filter for m_t is exactly equal to $\hat{m}_t$ of (49*) in view of (47*) and (48*), i.e.,

$$\hat{m}_t = a_0 + a_1 i_t + a_2 m_{t-1} + \sigma_n(r_t - b_0 - a_0 b_1 - a_2 b_1 m_{t-1} - (b_2 + a_1 b_1) i_t). \tag{56*}$$

To express the optimal monetary policy, we put $\hat{m}_t$ to the target value m^* in (56*), yielding a function linking the interest rate i_t and reserves r_t deterministically as given in LeRoy and Waud (1977, eq. (14)). In this case, the variance of m_t conditional on r_t (and i_t) becomes

$$\begin{aligned} E(m_t - \hat{m}_t)^2 &= E(u_{1t} - E(u_{1t} \mid u_{2t} + b_1 u_{1t}))^2 \\ &= \sigma_1^2 - 2b_1\sigma_1^2\sigma_n + (b_1^2\sigma_1^2 + \sigma_2^2)\sigma_n^2 \\ &= \sigma_1^2 - \frac{b_1^2\sigma_1^4}{b_1^2\sigma_1^2 + \sigma_2^2}, \end{aligned} \tag{57*}$$

which gives the loss under the optimum combination policy.

On the other hand, before observation of r_t (i.e., before the current realizations of u_{1t} and u_{2t} are taken into account), the estimates of m_t and r_t are derived by applying the Kalman predictor formula (37) on the assumption that the initial value $x(t-1)$ is known and nonrandom. Then, since $\hat{x}(t-1 \mid t-2)$ is set at $x(t-1)$, and in view of (51*), the formula (37) reduces to

$$\hat{x}(t \mid t-1) = Ax(t-1) + Bv(t) + k \tag{58*}$$

or, equivalently,

$$\hat{m}_{t \mid t-1} = a_0 + a_1 i_t + a_2 m_{t-1} \tag{59*a}$$

$$\hat{r}_{t \mid t-1} = b_0 + a_0 b_1 + a_2 b_1 m_{t-1} + (b_2 + a_1 b_1) i_t, \tag{59*b}$$

with covariance matrix of $x(t)$ given by $S(t)$ in (54*). Hence the optimal feedback rule is obtained as (60*), simply by putting $\hat{m}_{t \mid t-1}$ in (59*a) to the target m^*.

$$i_t = \frac{1}{a_1}(m^* - a_0 - a_2 m_{t-1}). \tag{60*}$$

For another interesting case, we refer the reader to Conrad and Corrado (1979), an application of the Kalman filter in a steady-state (i.e., in the case that $N(t)$ in our formula (48) is time-invariant) to revisions in retail sales estimates.

References

Anderson, B. D. O., and Moore, J. B., *Optimal Filtering*. Englewood Cliffs, NJ: Prentice-Hall, 1979.

Athans, M., "The importance of Kalman filtering methods for economic systems," *Annals of Economic and Social Measurements*, vol. 3, pp. 49–64, 1974.

Bertsekas, D. P., *Dynamic Programming and Stochastic Control*. New York: Academic Press, 1976.

Chow, G. C., *Analysis and Control of Dynamic Economic Systems*. New York: Wiley, 1975.

Conrad, W., and Corrado, C., "Application of the Kalman filter to revisions in monthly retail sales estimates," *J. Economic Dynamics and Control*, vol. 1, pp. 177–198, 1979.

Deyst, J. J., Jr., and Price, C. F., "Conditions for asymptotic stability of the discrete minimum-variance linear estimator," *IEEE Trans. Automatic Control*, vol. AC-13, pp. 702–705, 1968.

Jazwinski, A. H., *Stochastic Processes and Filtering Theory*. New York: Academic Press, 1970.

Kalman, R. E., "A new approach to linear filtering and prediction problems," *J. Basic Engineering, Trans. ASME (Series D)*, vol. 82, pp. 35–45, 1960.

Lazaridis, A., "Application of filtering methods in econometrics," *International J. Systems Science*, vol. 11, pp. 1315–1325, 1980.

LeRoy, S. F., and Waud, R. N., "Applications of the Kalman filter in short-run monetary control," *International Economic Review*, vol. 18, pp. 195–207, 1977.

Meditch, J. S., *Stochastic Optimal Linear Estimation and Control*. New York: McGraw-Hill, 1969.

Morf, M., Sidhu, G. S., and Kailath, T., "Some new algorithms for recursive estimation in constant, linear, discrete-time systems," *IEEE Trans. Automatic Control*, vol. AC-19, pp. 315–323, 1974.

Murata, Y., *Mathematics for Stability and Optimization of Economic Systems*. New York: Academic Press, 1977.

Pagan, A., "A note on the extraction of components from time series," *Econometrica*, vol. 43, pp. 163–168, 1975.

Rhodes, I. B., "A tutorial introduction to estimation and filtering," *IEEE Trans. Automatic Control*, vol. AC-16, pp. 688–706, 1971.

Taylor, L., "The existence of optimal distributed lags," *Review of Economic Studies*, vol. 37, pp. 95–106, 1970.

Tse, E., and Athans, M., "Optimal minimal-order observer-estimators for discrete linear time-varying systems," *IEEE Trans. Automatic Control*, vol. AC-15, pp. 416–426, 1970.

Vishwakarma, K. P., *Macro-Economic Regulation*. Rotterdam: Rotterdam University Press, 1974.

Yoshikawa, T., "Minimal-order optimal filters for discrete-time linear stochastic systems," *International J. Control*, vol. 21, pp. 1–19, 1975.

CHAPTER 5

Optimal Control of Linear Stochastic Discrete-Time Systems

In reality, any economic phenomenon occurs in an uncertain environment. We now consider optimal control for such dynamic systems. General rules are established for finite time-horizon optimal control problems of linear discrete-time systems with additive random disturbances both in perfect information cases (Section 5.1) and in imperfect information cases (Section 5.2). We also derive the optimal control rules for linear discrete-time systems with stochastic coefficients as well as additive disturbances (Section 5.3). Our approach is based primarily on the optimality principle in dynamic programming, except the end of Section 5.3 where we comment on Lagrange multiplier methods applicable to an infinite horizon problem. Some stochastic optimal control rules are found to be the same as for the corresponding nonstochastic systems in which additive random disturbances are suppressed. This is called the certainty equivalence, and we shall discuss the principle further in relation to Theil's strategy in Section 5.4. Finally, in Section 5.5, macroeconomic applications of our control rules will be presented together with other related control methodologies.

5.1. Controllers for Linear Systems with Additive Disturbances

Our concern in this section is to establish optimal control rules for linear discrete-time systems with additive random disturbances in perfect information cases. At first, we concentrate on a minimization problem for finite time horizon; in particular, we try to minimize the expected value of quadratic cost function:

$$J(x_0) \equiv \mathop{E}_{\xi(t);\, t=1,\ldots,\beta} \left\{ x^T(\beta)\Gamma x(\beta) + \sum_{t=1}^{\beta} \left(x^T(t-1)\Xi x(t-1) + v^T(t)\Phi v(t) \right) \right\} \tag{1}$$

with respect to instruments $v(1), \ldots, v(\beta)$, subject to the following state-space form of linear system:

$$x(t) = Ax(t-1) + Bv(t) + \xi(t), \qquad t = 1, 2, \ldots, \beta \tag{2}$$

for finite time-horizon β, with given initial value $x(0) = x_0$, where x is a state n vector, v is a control m vector, and ξ is a time-independent random n vector (not necessarily Gaussian) having zero mean and a finite constant covariance matrix R, i.e.,

$$E\xi(t) = 0, \tag{3a}$$

$$\operatorname{cov}(\xi(t), \xi(s)) = R\delta_{st}, \tag{3b}$$

where $\delta_{tt} = 1$ and $\delta_{st} = 0$ for $s \neq t$, that is, δ is the Kronecker delta. In the cost function (1), we assume as usual that Γ, Ξ, and Φ are constant positive semidefinite symmetric matrices with Φ being positive definite. (Cf. Åström (1970, ch. 8), and Meier *et al.* (1971).)

Objective (1), with (2) taken into consideration, is expressed as

$$J(x_0) = \mathop{E}_{\xi(t);\, t=1,\ldots,\beta} \left\{ \sum_{t=1}^{\beta} g_t(x(t-1), v(t), \xi(t)) \right\} + x_0^T \Xi x_0 \tag{1$'$}$$

where g_t $(t = 1, 2, \ldots, \beta - 1)$ and g_β are defined by (4a) and (4b) below, respectively.

$$g_t(x(t-1), v(t), \xi(t)) \equiv (Ax(t-1) + Bv(t) + \xi(t))^T \Xi \times (Ax(t-1) + Bv(t) + \xi(t)) + v^T(t)\Phi v(t), \tag{4a}$$

$$g_\beta(x(\beta-1), v(\beta), \xi(\beta)) \equiv (Ax(\beta-1) + Bv(\beta) + \xi(\beta))^T \Gamma \times (Ax(\beta-1) + Bv(\beta) + \xi(\beta)) + v^T(\beta)\Phi v(\beta). \tag{4b}$$

Let $J^*(x_0)$ be the optimal value of the expected cost $J(x_0)$ in (1′). Here we note that the probability measure characterizing $\xi(t)$ does not depend on prior values of disturbances $\xi(1), \ldots, \xi(t-1)$, and that control vector $v(t)$ is dependent on all the future optimization programs in view of the optimality principle in dynamic programming. Thus $J^*(x_0)$, the optimal

$J(x_0)$ in (1′), can be written

$$J^*(x_0) = \min_{v(1)}\Big[\mathop{E}_{\xi(1)}\Big\{ g_1(x_0, v(1), \xi(1)) + \min_{v(2)}\Big[\mathop{E}_{\xi(2)}\Big\{ g_2(x(1), v(2), \xi(2)) + \cdots$$

$$+ \min_{v(\beta)}\Big[\mathop{E}_{\xi(\beta)} g_\beta(x(\beta-1), v(\beta), \xi(\beta))\Big]\Big\}\Big]\cdots\Big\}\Big] + x_0^T\Xi x_0, \quad (5)$$

where the expectation E over $\xi(t)$ is conditional on $x(t-1)$ and $v(t)$, $t = 1, \ldots, \beta$.

Defining

$$J(\beta, \beta) \equiv \mathop{E}_{\xi(\beta)} g_\beta(x(\beta-1), v(\beta), \xi(\beta)), \quad (6a)$$

$$J(\beta, \beta-1) \equiv \mathop{E}_{\xi(\beta-1)}\Big\{ g_{\beta-1}(x(\beta-2), v(\beta-1), \xi(\beta-1)) + \min_{v(\beta)} J(\beta, \beta)\Big\}, \quad (6b)$$

$$\vdots$$

$$J(\beta, t) \equiv \mathop{E}_{\xi(t)}\Big\{ g_t(x(t-1), v(t), \xi(t)) + \min_{v(t+1)} J(\beta, t+1)\Big\}, \quad (6c)$$

$$\vdots$$

$$J(\beta, 1) \equiv \mathop{E}_{\xi(1)}\Big\{ g_1(x_0, v(1), \xi(1)) + \min_{v(2)} J(\beta, 2)\Big\} + x_0^T\Xi x_0, \quad (6d)$$

we apply Bellman's principle of optimality in dynamic programming, with (4) taken into consideration. Thus, in the first place, the differential of $J(\beta, \beta)$ with respect to $v(\beta)$ is set equal to zero, i.e.,

$$0 = \partial J(\beta, \beta)/\partial v(\beta)$$
$$= 2\big(B^T\Gamma B v(\beta) + B^T\Gamma A x(\beta-1) + \Phi v(\beta)\big) \quad (7a)$$

since, in view of (3a),

$$\mathop{E}_{\xi(\beta)}\big\{(Ax(\beta-1) + Bv(\beta) + \xi(\beta))^T\Gamma(Ax(\beta-1) + Bv(\beta) + \xi(\beta))\big\}$$
$$= (Ax(\beta-1) + Bv(\beta))^T\Gamma(Ax(\beta-1) + Bv(\beta))$$
$$+ E\big(\xi^T(\beta)\Gamma\xi(\beta)\big)$$

and since

$$E\big(\xi^T(\beta)\Gamma\xi(\beta)\big) = (E\xi(\beta))^T\Gamma(E\xi(\beta)) + \mathrm{tr}(\Gamma R) \quad (8)$$

in view of the following lemma.

Lemma 1. *Let M be an $n \times n$ constant matrix, and let y and z be random column n vectors having a finite covariance matrix $\mathrm{cov}(y, z)$. Then*

$$E(z^T M y) = (Ez)^T M(Ey) + \mathrm{tr}(M\,\mathrm{cov}(y, z)).$$

PROOF.

$$\begin{aligned}\operatorname{tr}(M\operatorname{cov}(y,z)) &= \operatorname{tr}\big(ME(y-Ey)(z-Ez)^T\big)\\ &= \operatorname{tr}\big(M\big[E(yz^T)-(Ey)(Ez)^T\big]\big)\\ &= \operatorname{tr}\big(ME(yz^T)\big)-\operatorname{tr}\big(M(Ey)(Ez)^T\big)\\ &= E\big(\operatorname{tr}(Myz^T)\big)-\operatorname{tr}\big((Ez)^TM(Ey)\big)\\ &= E\big(\operatorname{tr}(z^TMy)\big)-(Ez)^TM(Ey)\\ &= E(z^TMy)-(Ez)^TM(Ey)\end{aligned}$$

□

From (7a), the optimal control is derived as

$$v(\beta) = -K(\beta)x(\beta-1), \tag{9a}$$

where

$$K(\beta) \equiv \big[\Phi + B^T\Gamma B\big]^{-1}B^T\Gamma A. \tag{10a}$$

Substituting (9a) into (6a), we have the minimal value of $J(\beta, \beta)$ as

$$\begin{aligned}J^*(\beta,\beta) &= \underset{\xi(\beta)}{E}\Big\{\big(\big[A-BK(\beta)\big]x(\beta-1)\\ &\qquad + \xi(\beta)\big)^T\Gamma\big(\big[A-BK(\beta)\big]x(\beta-1)+\xi(\beta)\big)\Big\}\\ &\quad + x^T(\beta-1)K^T(\beta)\Phi K(\beta)x(\beta-1)\\ &= x^T(\beta-1)\Big[\big[A-BK(\beta)\big]^T\Gamma\big[A-BK(\beta)\big]\\ &\qquad + K^T(\beta)\Phi K(\beta)\Big]x(\beta-1)+\operatorname{tr}(\Gamma R)\end{aligned} \tag{6'a}$$

by virtue of (8).

Second, the differential of $J(\beta, \beta-1)$ with respect to $v(\beta-1)$ is equalized to zero. In view of (6b) and (6′a)

$$\begin{aligned}J(\beta,\beta-1) &= \underset{\xi(\beta-1)}{E}\Big\{\big(Ax(\beta-2)+Bv(\beta-1)+\xi(\beta-1)\big)^T\Xi\\ &\qquad \times\big(Ax(\beta-2)+Bv(\beta-1)+\xi(\beta-1)\big)+J^*(\beta,\beta)\Big\}\\ &\quad + v^T(\beta-1)\Phi v(\beta-1)\\ &= \underset{\xi(\beta-1)}{E}\Big\{\big(Ax(\beta-2)+Bv(\beta-1)+\xi(\beta-1)\big)^TS(\beta-1)\\ &\qquad \times\big(Ax(\beta-2)+Bv(\beta-1)+\xi(\beta-1)\big)\Big\}\\ &\quad + \operatorname{tr}(\Gamma R)+v^T(\beta-1)\Phi v(\beta-1)\\ &= \big(Ax(\beta-2)+Bv(\beta-1)\big)^TS(\beta-1)\\ &\quad \times\big(Ax(\beta-2)+Bv(\beta-1)\big)+\operatorname{tr}(S(\beta-1)R)+\operatorname{tr}(\Gamma R)\\ &\quad + v^T(\beta-1)\Phi v(\beta-1),\end{aligned} \tag{6'b}$$

where

$$S(\beta-1) \equiv [A - BK(\beta)]^T\Gamma[A - BK(\beta)] + K^T(\beta)\Phi K(\beta) + \Xi$$
$$= A^T\Gamma(A - BK(\beta)) + \Xi$$

due to (10a). Thus

$$0 = \partial J(\beta, \beta-1)/\partial v(\beta-1)$$
$$= 2\big(B^TS(\beta-1)Bv(\beta-1) + B^TS(\beta-1)Ax(\beta-2) + \Phi v(\beta-1)\big), \tag{7b}$$

from which follows the optimal value of $v(\beta-1)$

$$v(\beta-1) = -K(\beta-1)x(\beta-2), \tag{9b}$$

where

$$K(\beta-1) \equiv \left[\Phi + B^TS(\beta-1)B\right]^{-1}B^TS(\beta-1)A. \tag{10b}$$

Substitution of (9b) into (6′b) yields the minimal value of $J(\beta, \beta-1)$ as

$$J^*(\beta, \beta-1) = x^T(\beta-2)\Big[(A - BK(\beta-1))^TS(\beta-1)(A - BK(\beta-1))$$
$$+ K^T(\beta-1)\Phi K(\beta-1)\Big]x(\beta-2)$$
$$+ \operatorname{tr}(\Gamma + S(\beta-1))R.$$

Proceeding in this manner, we get a general rule.

Theorem 1. *The optimal control of the finite time-horizon minimization problem that objective* (1) *be minimized for system* (2), *is given by*

$$v(t) = -K(t)x(t-1) \qquad \textit{for} \quad t = 1, \ldots, \beta, \tag{11}$$

where

$$K(t) \equiv \left[\Phi + B^TS(t)B\right]^{-1}B^TS(t)A \tag{12a}$$

$$S(t-1) = A^TS(t)\left[A - BK(t)\right] + \Xi, \qquad t = 2, 3, \ldots, \beta$$
$$= A^T\Big[S(t) - S(t)B\left[\Phi + B^TS(t)B\right]^{-1}B^TS(t)\Big]A + \Xi \tag{12b}$$

$$S(\beta) = \Gamma. \tag{12c}$$

The associated minimal value of the cost functional is

$$J^*(x_0) = \min_{v(1)} J(\beta, 1)$$
$$= x_0^TS(0)x_0 + \operatorname{tr}(S(1) + \cdots + S(\beta))R. \tag{13}$$

Remark. It is important to note that the optimal control law (11) for the stochastic system (2) is the same as for the corresponding deterministic system in which stochastic term $\xi(t)$ disappears and instead its expectation $E\xi(t)$ appears. This property of the present linear control system with an additive stochastic term is usually called the *certainty equivalence principle.* (See Section 5.4 for further discussions on the principle.)

Corollary. *Similarly, we can show that when our system contains a non-stochastic exogenous term in addition to the original ones in system* (2), *i.e.*,

$$x(t) = Ax(t-1) + Bv(t) + c(t) + \xi(t), \tag{2'}$$

where $c(t)$ is a vector of nonrandom exogenous variables, the certainty equivalence principle still holds, and hence the optimal control is given by (cf. Theorem 14 in Section 2.2)

$$v(t) = -K(t)x(t-1) - k(t), \qquad t = 1, \ldots, \beta \tag{11'}$$

where $K(t)$ is that of (12*a*),

$$k(t) \equiv \left[B^T S(t) B + \Phi\right]^{-1} B^T \Bigg\{ S(t)c(t) + L(t+1)S(t+1)c(t+1) + L(t+1)L(t+2)S(t+2)c(t+2) + \ldots + \prod_{\tau=1}^{\beta-t} L(\beta-\tau+1)S(\beta)c(\beta)\Bigg\}, \tag{12'a}$$

$S(t)$ and $S(\beta)$ are those of (12*b*) *and* (12*c*), *respectively, and*

$$L(t) \equiv \left[A - BK(t)\right]^T, \qquad t = 2, 3, \ldots, \beta. \tag{12'b}$$

Next, we consider the infinite time-horizon minimization problem with additive disturbances; that is, we minimize the expected value of cost functional

$$J^{\dagger}(x_0) \equiv E\sum_{t=1}^{\infty}\left(x^T(t-1)\Xi x(t-1) + v^T(t)\Phi v(t)\right) \tag{1†}$$

for given $x(0) = x_0$, with respect to $v(t)$ subject to system (2) for $t = 1, 2, \cdots, \infty$. The presupposed properties of additive disturbances in (2), as well as the other notations, are the same as for the finite time-horizon case.

As Theorem 15 in Section 2.2 above suggests, by virtue of the certainty equivalence principle, if the deterministic system corresponding to (2) is state controllable, we have the optimal control law to the infinite time-horizon problem as follows, in contrast with Theorem 1. (Cf. Bertsekas (1976, Sec. 3.1) for this problem.)

Theorem 2. *The optimal control to the infinite time-horizon problem of minimizing objective* ($1^\dagger$) *for system* (2), *is given by*

$$v(t) = -Kx(t-1) \qquad \textit{for} \quad t = 1,2,\ldots \tag{11†}$$

where

$$K \equiv \left[B^TSB + \Phi\right]^{-1} B^TSA \tag{12†a}$$

$$S = A^TS\left[A - BK\right] + \Xi. \tag{12†b}$$

Furthermore, if system (2) *is replaced by system* (2′), *the present problem is solved for the following optimal control rule.*

$$v(t) = -Kx(t-1) - \bar{k}(t) \qquad \textit{for} \quad t = 1,2,\ldots \tag{11††}$$

where K is that of ($12^\dagger$a),

$$\bar{k}(t) \equiv \left[B^TSB + \Phi\right]^{-1} B^T\left\{Sc(t) + LSc(t+1) + L^2Sc(t+2) + \ldots\right\}, \tag{12††a}$$

$$L \equiv \left[A - BK\right]^T, \qquad (\textit{note that } L^\infty = 0) \tag{12††b}$$

and S is that of ($12^\dagger$b).

Concluding this section, we extend control rule (11) to a distributed-lag system case.

Application 1. We apply the optimal control rule (11) to the following distributed-lag system

$$x(t) = \sum_{i=1}^{k} A_i x(t-i) + \sum_{j=0}^{h} B_j v(t-j) + \xi(t) \tag{1*}$$

with the cost functional (2*) to be minimized

$$J = \sum_{t=1}^{\beta}\left\{\sum_{i=1}^{k} x^T(t-i)\Xi_i x(t-i) + \sum_{j=1}^{h} v^T(t-j)\Xi_{k+j}v(t-j) + v^T(t)\Phi v(t)\right\} + x^T(\beta)\Gamma_0 x(\beta) \tag{2*}$$

where $x(t)$ and $v(t)$ are the state n vector and control m vector, respectively, in period t; $\xi(t)$ is a stochastic term obeying assumptions (3); A_i, B_j are constant matrices of appropriate dimensions; and Ξ_i ($i = 1,2,\ldots,$ $k+h$), Φ, Γ_0 are all constant positive definite matrices.

Equations (1*) and (2*) are rewritten as

$$\tilde{x}(t) = \tilde{A}\tilde{x}(t-1) + \tilde{B}v(t) + \tilde{D}\xi(t), \tag{1**}$$

and

$$J = \sum_{t=1}^{\beta} \left\{ \tilde{x}^T(t-1)\Xi\tilde{x}(t-1) + v^T(t)\Phi v(t) \right\} + \tilde{x}^T(\beta)\Gamma\tilde{x}(\beta), \quad (2^{**})$$

where

$$\tilde{A} \equiv \begin{bmatrix} 0 & I_m & & 0 & 0 & \cdots & \cdots & 0 \\ \vdots & \ddots & \ddots & & & & & \vdots \\ \vdots & & 0 & I_m & 0 & \cdots & \cdots & 0 \\ 0 & \cdots & 0 & 0 & 0 & \cdots & \cdots & 0 \\ 0 & \cdots & \cdots & 0 & 0 & I_n & & 0 \\ \vdots & & & & & \ddots & \ddots & \\ 0 & \cdots & \cdots & 0 & 0 & \cdots & 0 & I_n \\ B_h & \cdots & \cdots & B_1 & A_k & \cdots & A_2 & A_1 \end{bmatrix}, \qquad \tilde{B} \equiv \begin{bmatrix} 0 \\ \vdots \\ 0 \\ I_m \\ 0 \\ \vdots \\ 0 \\ B_0 \end{bmatrix},$$

$$\tilde{x}(t) \equiv \begin{bmatrix} v(t-h+1) \\ \vdots \\ v(t-1) \\ v(t) \\ x(t-k+1) \\ \vdots \\ x(t-1) \\ x(t) \end{bmatrix}, \qquad \tilde{D} \equiv \begin{bmatrix} 0 \\ \vdots \\ \vdots \\ \vdots \\ \vdots \\ 0 \\ I_n \end{bmatrix} \qquad \Gamma \equiv \begin{bmatrix} 0 & & & \\ & \ddots & & \\ & & 0 & \\ & & & \Gamma_0 \end{bmatrix},$$

and

$$\Xi \equiv \operatorname{diag}(\Xi_{k+h}, \ldots, \Xi_{k+1}, \Xi_k, \ldots, \Xi_1). \quad (3^*)$$

Define

$$K(t) \equiv \left[\tilde{B}^T S(t) \tilde{B} + \Phi \right]^{-1} \tilde{B}^T S(t) \tilde{A}, \quad (4^*)$$

$$S(t) = \left[\tilde{A} - \tilde{B}K(t) \right]^T S(t+1) \left[\tilde{A} - \tilde{B}K(t) \right] + K^T(t)\Phi K(t) + \Xi \quad (5^*)$$

for $t = 1, 2, \ldots, \beta - 1$; and

$$S(\beta) = \Gamma. \quad (6^*)$$

Taking account of the definition of $\tilde{A}$, $\tilde{B}$ and of the dimension of $K(t)$, we partition $K(t)$ as follows.

$$K(t) = \left[K_{Bh}(t), K_{Bh-1}(t), \ldots, K_{B1}(t), K_{Ak}(t), K_{Ak-1}(t), \ldots, K_{A1}(t) \right], \quad (7^*)$$

where each $K_{Bj}(t)$ $(j = 1, \ldots, h)$ has dimension $m \times m$, and each $K_{Ai}(t)$

$(i = 1, \ldots, k)$ has dimension $m \times n$, defined as

$$K_{Bh}(t) \equiv \left[\tilde{B}^T S(t)\tilde{B} + \Phi\right]^{-1} U(t)B_h,$$

$$K_{Bh-1}(t) \equiv \left[\tilde{B}^T S(t)\tilde{B} + \Phi\right]^{-1}\left[U(t)B_{h-1} + S_{h1}(t) + B_0^T S_{k+h,1}(t)\right],$$

$$\vdots$$

$$K_{B1}(t) \equiv \left[\tilde{B}^T S(t)\tilde{B} + \Phi\right]^{-1}\left[U(t)B_1 + S_{h,h-1}(t) + B_0^T S_{k+h,h-1}(t)\right]; \quad (8^*)$$

$$K_{Ak}(t) \equiv \left[\tilde{B}^T S(t)\tilde{B} + \Phi\right]^{-1} U(t)A_k,$$

$$K_{Ak-1}(t) \equiv \left[\tilde{B}^T S(t)\tilde{B} + \Phi\right]^{-1}\left[U(t)A_{k-1} + S_{h,h+1}(t) + B_0^T S_{k+h,h+1}(t)\right],$$

$$\vdots$$

$$K_{A1}(t) \equiv \left[\tilde{B}^T S(t)\tilde{B} + \Phi\right]^{-1}\left[U(t)A_1 + S_{h,k+h-1}(t) + B_0^T S_{k+h,k+h-1}(t)\right],$$

in which

$$U(t) \equiv S_{h,k+h}(t) + B_0^T S_{k+h,k+h}(t).$$

The following $m \times (hm + kn)$ and $n \times (hm + kn)$ matrices

$$\left[S_{h1}(t), \ldots, S_{hh}(t), S_{h,h+1}(t), \ldots, S_{h,k+h}(t)\right],$$

$$\left[S_{k+h,1}(t), \ldots, S_{k+h,h}(t), S_{k+h,h+1}(t), \ldots, S_{k+h,k+h}(t)\right]$$

constitute, respectively, the hth row block (i.e., rows $(h-1)m + 1 \sim hm$) and the $(k+h)$th row block (i.e., the last n rows) of $S(t)$. Applying the optimal control rule (11) with $x(t-1)$ replaced by $\tilde{x}(t-1)$ in (1**), therefore, yields

$$v(t) = -K(t)\tilde{x}(t-1) \quad (9^*)$$

or, equivalently, in view of (7*)

$$v(t) = -\sum_{j=1}^{h} K_{Bj}(t)v(t-j) - \sum_{i=1}^{k} K_{Ai}(t)x(t-i). \quad (9^{**})$$

5.2. Controller in an Imperfect Information Case

In this section, we are concerned with optimal control of linear discrete-time systems with additive disturbances in an imperfect state information case. Thus we consider the following system in a state-space form similar to that in Kalman estimation (Section 4.1): for $t = 1, \ldots, \beta$

$$x(t) = Ax(t-1) + Bv(t) + \xi(t), \quad (14a)$$

$$y(t) = Cx(t) + \zeta(t), \qquad \text{with} \quad C \equiv [I_r, 0] \quad (14b)$$

where y and ζ are r vectors of output variables and time-independent non-Gaussian observation errors, respectively. (The other notations are those of (2) in Section 5.1.) State vector $x(t)$ is assumed to be partially observed only through a noise-spoiled output $y(t)$ of dimension r ($< n$). The corresponding cost function will become

$$J = E\left\{ y^T(\beta)\Gamma y(\beta) + \sum_{t=1}^{\beta} \left(y^T(t-1)\Xi y(t-1) + v^T(t)\Phi v(t)\right)\right\}, \quad (15)$$

where Γ, Ξ are positive semidefinite, and Φ is positive definite.

Partitioning n vector x as

$$x = \begin{pmatrix} x_1 \\ x_2 \end{pmatrix}, \quad (16)$$

where x_1 has dimension r and is assumed to correspond to output y, and x_2 is an unobservable $(n-r)$ vector. As in (16), we partition matrices A and B, and disturbance term ξ as

$$A = \begin{pmatrix} A_{11} & A_{12} \\ A_{21} & A_{22} \end{pmatrix}, \qquad B = \begin{pmatrix} B_{11} \\ B_{22} \end{pmatrix}, \qquad \xi = \begin{pmatrix} \xi_1 \\ \xi_2 \end{pmatrix}. \quad (17)$$

Defining

$$\tilde{x}(t) \equiv \begin{pmatrix} x_1(t) + \zeta(t) \\ x_2(t) \end{pmatrix} = \begin{pmatrix} y(t) \\ x_2(t) \end{pmatrix}, \qquad \tilde{\xi}(t) \equiv \begin{pmatrix} \xi_1(t) + \zeta(t) - A_{11}\zeta(t-1) \\ \xi_2(t) - A_{21}\zeta(t-1) \end{pmatrix}, \quad (18)$$

we can rewrite system (14) as

$$\tilde{x}(t) = A\tilde{x}(t-1) + Bv(t) + \tilde{\xi}(t), \quad (14')$$

and, accordingly, cost function (15) is expressed as

$$J = E\left\{ \tilde{x}^T(\beta)\tilde{\Gamma}\tilde{x}(\beta) + \sum_{t=1}^{\beta} \left(\tilde{x}^T(t-1)\tilde{\Xi}\tilde{x}(t-1) + v^T(t)\Phi v(t)\right)\right\}, \quad (15')$$

where

$$\tilde{\Gamma} \equiv \begin{pmatrix} \Gamma & 0 \\ 0 & 0 \end{pmatrix}, \qquad \tilde{\Xi} \equiv \begin{pmatrix} \Xi & 0 \\ 0 & 0 \end{pmatrix}.$$

As ζ and ξ are non-Gaussian random vectors having zero means and finite covariance matrices, so $\tilde{\xi}$ is a non-Gaussian random vector having the same properties. Hence we may consider, from the outset, the transformed system

$$x(t) = Ax(t-1) + Bv(t) + \xi(t), \qquad \text{for} \quad t = 1, \ldots, \beta \quad (19)$$

with the cost functional to be minimized

$$J = E\left\{ x^T(\beta)\Gamma x(\beta) + \sum_{t=1}^{\beta} \left(x^T(t-1)\Xi x(t-1) + v^T(t)\Phi v(t)\right)\right\}, \quad (20)$$

where x is a state n vector, v is a control m vector, and ξ is assumed to be a non-Gaussian random n vector having the following properties:

$$E\xi(t) = 0 \qquad \text{for all } t, \quad (21a)$$

$$\operatorname{cov}(\xi(t), \xi(s)) = R\delta_{st} \qquad (\delta_{tt} = 1,\ \delta_{st} = 0\ (s \neq t)). \quad (21b)$$

Using partition (17), we divide system (19) into two subsystems:

$$x_1(t) = A_{11}x_1(t-1) + A_{12}x_2(t-1) + B_{11}v(t) + \xi_1(t), \tag{22a}$$

$$x_2(t) = A_{21}x_1(t-1) + A_{22}x_2(t-1) + B_{22}v(t) + \xi_2(t). \tag{22b}$$

While $x_1(t)$ corresponds directly to $y(t)$, information about $x_2(t)$ must be searched for. Subsystems (22) show that $x_2(t)$ is conditional on $x_1(0)$, $x_1(1), \ldots, x_1(t-1)$ and $v(1), \ldots, v(t)$. We find the distribution function of $x_2(t)$ when $x_1(t)$ is given, following the method developed by Root (1969). (Cf. Aoki (1967) for related topics.)

Let us put

$$a = x_1(t),$$

$$F^t = \text{the conditional distribution of } x_2(t), \text{ given } x_1(0), x_1(1), \ldots, x_1(t) = a, \text{ and } v(1), \ldots, v(t).$$

Assume that distribution functions of $\xi(t)$ and $x(0)$ admit ordinary densities and have finite covariance matrices. The assumption implies that F^t admits an ordinary density and has a finite covariance matrix:

$$Q(F^t) \equiv E(x_2(t) - m(t))(x_2(t) - m(t))^T, \tag{23}$$

where

$$m(t) \equiv E\{x_2(t) : F^t\}. \tag{24}$$

In this (24), the term in the braces "$x_2(t): F^t$" denotes $x_2(t)$ whose distribution function is F^t. In the following, we employ this convention. Note that

$$\bar{x}(t) \equiv E\{x(t) \mid x_1(0), \ldots, x_1(t) = a, v(1), \ldots, v(t)\}$$
$$= \begin{pmatrix} a \\ m(t) \end{pmatrix}. \tag{25}$$

Also, writing for any vector b

$$F_b^t(t) \equiv F^t(x - b), \tag{26}$$

we know

$$Q(F^t) = Q(F_b^t) \qquad \text{for any } b. \tag{23$'$}$$

The marginal distribution function of $x_1(t)$ is

$$H^t(\eta, F^{t-1}, a, v) \equiv \Pr\left[x_1(t) \leqslant \eta \mid x_2(t-1) : F^{t-1}, x_1(t-1) = a, v(t) = v\right]$$
$$= \Pr\left[A_{12}x_2(t-1) + \xi_1(t) \leqslant \eta - A_{11}a - B_{11}v \mid x_2(t-1): F^{t-1}, x_1(t-1) = a, v(t) = v\right]$$
$$= \Pr\left[A_{12}x_2(t-1) + \xi_1(t) \leqslant \eta - A_{11}a + A_{12}b - B_{11}v \mid x_2(t-1) : F_b^{t-1}, x_1(t-1) = a, v(t) = v\right]$$
$$\equiv H^{*t}(\eta - A_{11}a + A_{12}b - B_{11}v, F_b^{t-1}), \tag{27}$$

and the joint distribution function of $\{x_1(t), x_2(t)\}$ is

$$\begin{aligned}
G^t(\mu, F^{t-1}, a, v) &\equiv \Pr\left[x(t) \leqslant \mu \mid x_2(t-1) : F^{t-1}, x_1(t-1) = a, v(t) = v\right] \\
&= \Pr\left[\begin{pmatrix} A_{12} \\ A_{22} \end{pmatrix} x_2(t-1) + \xi(t) \leqslant \mu - \begin{pmatrix} A_{11} \\ A_{21} \end{pmatrix} a - Bv \right| \\
&\qquad \left. x_2(t-1) : F^{t-1}, x_1(t-1) = a, v(t) = v\right] \\
&= \Pr\left[\begin{pmatrix} A_{12} \\ A_{22} \end{pmatrix} x_2(t-1) + \xi(t) \leqslant \mu - \begin{pmatrix} A_{11} \\ A_{21} \end{pmatrix} a + \begin{pmatrix} A_{12} \\ A_{22} \end{pmatrix} b \right. \\
&\qquad \left. - Bv \middle| x_2(t-1) : F_b^{t-1}, x_1(t-1) = a, v(t) = v\right] \\
&\equiv G^{*t}\left[\mu - \begin{pmatrix} A_{11} \\ A_{21} \end{pmatrix} a + \begin{pmatrix} A_{12} \\ A_{22} \end{pmatrix} b - Bv, F_b^{t-1}\right]. \qquad (28)
\end{aligned}$$

If F is a distribution function with an ordinary density, we shall denote the corresponding density function by its lower-case letter f. Using this convention, the distribution function of $x_2(t)$, given $x_1(t) = c$, is written as

$$\begin{aligned}
&D^t(w, e, c, F^{t-1}, a, v) \\
&\quad \equiv \Pr\left[x_2(t) + e \leqslant w \mid x_2(t-1) : F^{t-1},\right. \\
&\qquad \left. x_1(t-1) = a, v(t) = v, x_1(t) = c\right] \\
&\quad = \int_{-\infty}^{w} \frac{g^{*t}\left[\begin{pmatrix} c \\ z - e \end{pmatrix} - \begin{pmatrix} A_{11} \\ A_{21} \end{pmatrix} a + \begin{pmatrix} A_{12} \\ A_{22} \end{pmatrix} b - Bv, F_b^{t-1}\right]}{h^{*t}(c - A_{11}a + A_{12}b - B_{11}v, F_b^{t-1})}\, dz \qquad (29)
\end{aligned}$$

for any b. Hence

$$\begin{aligned}
&D^t(w, A_{22}b - A_{21}a - B_{22}v, c, F^{t-1}, a, v) \\
&\quad = \int_{-\infty}^{w} \frac{g^{*t}\left[\begin{pmatrix} c + A_{12}b - A_{11}a - B_{11}v \\ z \end{pmatrix}, F_b^{t-1}\right]}{h^{*t}(c - A_{11}a + A_{12}b - B_{11}v, F_b^{t-1})}\, dz \\
&\quad \equiv D^{*t}(w, c - A_{11}a + A_{12}b - B_{11}v, F_b^{t-1}). \qquad (29')
\end{aligned}$$

Now we are in a position to establish the optimal control law for the present imperfect state information case.

Theorem 3 (Separation Theorem). *The optimal control of system* (19) *with cost functional* (20) *to be minimized is given by*

$$v(t) = -K(t)\bar{x}(t-1) \qquad \text{for} \quad t = 1, \ldots, \beta \qquad (30)$$

where $\bar{x}(t)$ *is that of* (25), $K(t)$ *is that of* (12a), *and the associated minimum expected cost from* $t+1$ *till* β *becomes*

$$J^{t+1}(a, F^t) = (a^T, m^T(t))S(t)\binom{a}{m(t)} + \lambda^{t+1}(F^t), \tag{31}$$

where $S(t)$ *is the positive semidefinite symmetric matrix defined in* (12b), *and*

$$\lambda^{t+1}(F^t) = \lambda^{t+1}(F_b^t) \qquad \text{for any } b \text{ and } F^t. \tag{32}$$

PROOF. The proof is inductive, proceeding backward in time from the final period β, by using a dynamic programming argument. In view of (20), by putting $x_1(\beta-1) = a$,

$$J^\beta(a, F^{\beta-1}) \equiv \min_{v(\beta)} E\{x^T(\beta)\Gamma x(\beta) + x^T(\beta-1)\Xi x(\beta-1) + v^T(\beta)\Phi v(\beta) \mid (a, F^{\beta-1})\},$$

in which

$$x(\beta) = Ax(\beta-1) + Bv(\beta) + \xi(\beta),$$

and (21a) are taken into consideration, we have

$$\begin{aligned} J^\beta(a, F^{\beta-1}) = \min_{v(\beta)} \{ & E(x^T(\beta-1)[A^T\Gamma A + \Xi]x(\beta-1)) \\ & + 2v^T(\beta)B^T\Gamma A\bar{x}(\beta-1) + v^T(\beta)[\Phi + B^T\Gamma B]v(\beta) \\ & + 2E(x^T(\beta-1)A^T\Gamma\xi(\beta)) \\ & + E(\xi^T(\beta)\Gamma\xi(\beta)) \mid (a, F^{\beta-1})\}. \end{aligned}$$

Differentiating the function in the brackets of this J^β with respect to $v(\beta)$ and equalizing the resultant to zero, we obtain

$$0 = B^T\Gamma A\bar{x}(\beta-1) + [\Phi + B^T\Gamma B]v(\beta),$$

whence the optimal value of $v(\beta)$ is derived as

$$v(\beta) = -K(\beta)\bar{x}(\beta-1), \tag{30$'$}$$

where

$$K(\beta) \equiv [\Phi + B^T\Gamma B]^{-1}B^T\Gamma A.$$

This $K(\beta)$ is the same as in (10a) of the preceding section for the perfect state information case. Substitution of (30′) into the above $J^\beta(a, F^{\beta-1})$ yields its minimum value

$$\begin{aligned} J^\beta(a, F^{\beta-1}) = \bar{x}^T(\beta-1)&\left[\Xi + A^T(\Gamma - \Gamma B[\Phi + B^T\Gamma B]^{-1}B^T\Gamma)A\right] \\ &\times \bar{x}(\beta-1) + \mathrm{tr}([\Xi + A^T\Gamma A]_{22}Q(F^{\beta-1})) + \mathrm{tr}(\Gamma R) \end{aligned} \tag{31$'$}$$

where we have taken account of Lemma 1, of (21b), of the fact that

$$E\big(x^T(\beta-1)A^T\Gamma\xi(\beta)\big)=0$$

since $x(\beta-1)$ is independent of $\xi(\beta)$, and of the fact that

$$\operatorname{cov}\left[\begin{pmatrix} a \\ x_2(\beta-1)\end{pmatrix},\begin{pmatrix} a \\ x_2(\beta-1)\end{pmatrix}\right]=\begin{pmatrix} 0 & 0 \\ 0 & Q(F^{\beta-1})\end{pmatrix}; \tag{33}$$

and $[\Xi + A^T\Gamma A]_{22}$ denotes the submatrix of $[\Xi + A^T\Gamma A]$ corresponding to $Q(F^{\beta-1})$. The matrix in the first term on the right-hand side of (31′) is positive semidefinite since

$$\begin{aligned}
&\Xi + A^T\Gamma A - A^T\Gamma B\big[\Phi + B^T\Gamma B\big]^{-1}B^T\Gamma A \\
&\quad = \Xi + A^T\Gamma\big[A - BK(\beta)\big] \\
&\quad = \Xi + \big[A - BK(\beta)\big]^T\Gamma\big[A - BK(\beta)\big] + K^T(\beta)\Phi K(\beta).
\end{aligned}\tag{34}$$

Therefore, the statement in the theorem holds true for $t=\beta-1$.

Next, assuming the statement holds for an arbitrary t, we shall show that it also holds for $t-1$. By putting $x_1(t-1)=c$,

$$\begin{aligned}
J^t(c,F^{t-1}) \equiv \min_{v(\tau);\,\tau=t,\ldots,\beta} E\Bigg\{ & x^T(\beta)\Gamma x(\beta) + \sum_{\tau=t}^{\beta}\big(x^T(\tau-1)\Xi x(\tau-1) \\
& + v^T(\tau)\Phi v(\tau)\,|\,(c,F^{t-1}\big)\Bigg\} \\
= \min_{v(t)} E\big[& x^T(t-1)\Xi x(t-1) + v^T(t)\Phi v(t)\,|\,(c,F^{t-1})\big] \\
+ \min_{v(t)} E\Bigg\{ & \min_{v(\tau);\,\tau=t+1,\ldots,\beta} E\Bigg[x^T(\beta)\Gamma x(\beta) \\
& + \sum_{\tau=t+1}^{\beta}\big(x^T(\tau-1)\Xi x(\tau-1) \\
& + v^T(\tau)\Phi v(\tau)\big)\,|\,(a,F^t)\Bigg] \\
& |a : H^t(\cdot,F^{t-1},c,v)\Bigg\},
\end{aligned}\tag{35}$$

where we set $v(t)=v$, $x_1(t)=a$. The first term on the right-hand side of (35) is reduced to

$$\min_v\big\{\bar{x}^T(t-1)\Xi\bar{x}(t-1) + \operatorname{tr}\big(\Xi_{22}Q(F^{t-1})\big) + v^T\Phi v\big\},$$

where Ξ_{22} stands for the submatrix of Ξ corresponding to $Q(F^{t-1})$. (Cf.

(33).) Its second term in the braces reduces to

$$\left[J^{t+1}(a, F^t) \mid a : H^t(\cdot, F^{t-1}, c, v)\right].$$

Hence

$$J^t(c, F^{t-1}) = \min_v \left\{ \bar{x}^T(t-1)\Xi\bar{x}(t-1) + \operatorname{tr}\left(\Xi_{22}Q(F^{t-1})\right) + v^T\Phi v \right.$$
$$\left. + E\left[J^{t+1}(a, F^t) \mid a : H^t(\cdot, F^{t-1}, c, v)\right]\right\}. \qquad (35')$$

In (35′), note that

$$F^t = D^t(\cdot, 0, a, F^{t-1}, c, v). \qquad (36)$$

By induction hypothesis, the last term in the braces on the right-hand side of (35′) is rewritten

$$E\left[(a^T, m^T(t))S(t)\begin{pmatrix} a \\ m(t) \end{pmatrix} + \lambda^{t+1}(F^t) \mid a : H^t(\cdot, F^{t-1}, c, v)\right].$$

We examine (35′) term by term. Because of (32), for any e

$$E\left[\lambda^{t+1}(F^t) \mid a : H^t(\cdot, F^{t-1}, c, v)\right]$$
$$= E\left[\lambda^{t+1}\left(D^t(\cdot, e, a, F^{t-1}, c, v)\right) \mid a : H^t\right]$$
$$= E\left[\lambda^{t+1}\left(D^{*t}(\cdot, z, F_b^{t-1})\right) \mid z : H^{*t}(z, F_b^{t-1})\right] \qquad \text{for any } b$$
$$= \int_{-\infty}^{\infty} \lambda^{t+1}\left(D^{*t}(\cdot, z, F_b^{t-1})\right) h^{*t}(z, F_b^{t-1})\,dz$$
$$\equiv \pi^t(F_b^{t-1}), \qquad (37)$$

where z stands for $a + A_{12}b - A_{11}c - B_{11}v$; and we note that

$$\pi^t(F^{t-1}) = \pi^t(F_b^{t-1}) \qquad \text{for any } b$$

and it is independent of v. Next, in view of Lemma 1,

$$E\left[(a^T, m^T(t))S(t)\begin{pmatrix} a \\ m(t) \end{pmatrix} \mid a : H^T\right]$$
$$= E\left\{\left[-\operatorname{tr}\left(S_{22}(t)Q(F^t)\right)\right.\right.$$
$$\left.\left. + E\left(x^T(t)S(t)x(t) \mid x_2(t) : F^t, x_1(t) = a\right)\right] \mid a : H^t\right\},$$

where S_{22} denotes the submatrix of S corresponding to $Q(F^t)$. (Cf. (33).) Since

$$E\left[Q(F^t) \mid x_2(t-1) : F_b^{t-1}, x_1(t-1) = c, v(t) = v\right]$$

is independent of c, v, and b, we may write

$$E\left[\operatorname{tr}\left(S_{22}(t)Q(F^t)\right) \mid a : H^t\right] = M^t(F^{t-1}), \qquad (38)$$

where

$$M^t(F^{t-1}) = M^t(F_b^{t-1}) \qquad \text{for any } b$$

and it is independent of v. Lastly, in view of (36)

$$L \equiv E\left\{E\left[x^T(t)S(t)x(t) \mid x_2(t) : F^t, x_1(t) = a\right] \mid a : H^t(\cdot, F^{t-1}, c, v)\right\}$$
$$= E\left\{x^T(t)S(t)x(t) \mid x_2(t-1) : F^{t-1}, x_1(t-1) = c,\, v(t) = v\right\},$$

with $x(t)$ replaced by $Ax(t-1) + Bv(t) + \xi(t)$. Thus

$$L = E\left\{x^T(t-1)A^TS(t)Ax(t-1) + 2v^TB^TS(t)Ax(t-1) + v^TB^TS(t)Bv\right.$$
$$\left. + \xi^T(t)S(t)\xi(t) \mid x_2(t-1) : F^{t-1}, x_1(t-1) = c, v\right\}. \tag{39}$$

The optimal control value of v is provided by differentiating the expectation function on the right-hand side of (35) with (39) taken into consideration and by setting the resultant equal to zero, i.e.,

$$0 = \left[\Phi + B^TS(t)B\right]v + B^TS(t)A\bar{x}(t-1),$$

and hence optimal v is given by

$$v = -K(t)\bar{x}(t-1), \tag{30''}$$

where

$$K(t) \equiv \left[\Phi + B^TS(t)B\right]^{-1}B^TS(t)A.$$

Substituting v of (30″) back into (35′), with (37), (38), and (39) taken into account, yields the minimal value of $J^t(c, F^{t-1})$, i.e.,

$$J^t(c, F^{t-1}) = \bar{x}^T(t-1)\left[\Xi + A^TS(t)\left[A - BK(t)\right]\right]\bar{x}(t-1)$$
$$+ \operatorname{tr}\left(\left(\Xi_{22} + \left[A^TS(t)A\right]_{22}\right)Q(F^{t-1})\right) + \operatorname{tr}(S(t)R)$$
$$+ \pi^t(F^{t-1}) - M^t(F^{t-1}). \tag{40}$$

It is easily seen that (40) is of the same form as (31). □

Theorem 3 allows us to estimate $\bar{x}(t)$ and optimal control separately and independently. Thus the theorem is termed the *separation theorem* in control theory. (Cf. Witsenhausen (1971).) It can be regarded as the stochastic counterpart of the separation principle of observer and controller established in Section 3.5.

Application 2. Consider the same distributed-lag system (1*) with cost function (2*) as in Application 1, but assume imperfect state information for contemporaneous state variables. Then following a similar argument to the previous application, we have optimal control rule as

$$v(t) = -\sum_{j=1}^{h} K_{Bj}(t)v(t-j) - \sum_{i=2}^{k} K_{Ai}(t)x(t-i) - K_{A1}(t)\bar{x}(t-1), \tag{10*}$$

where $K_{Bj}(t)$, $K_{Ai}(t)$ are those of (8*) and $\bar{x}(t-1)$ is one of the Kalman estimators of $x(t-1)$.

Application 3. We want to apply the separation theorem (Theorem 3) to the optimal open market strategy problem by Kareken *et al.* (1973). In their linear economic system, discussed shortly, which holds over days, three variables exist: a nominal rate of interest R_t, the central bank's asset portfolio (stock of reserves) P_t, and nominal GNP Y_t for day t. They assume that both the interest rate and the asset portfolio are observed by the central bank without lapse of time, but that daily GNP cannot be observed without lapse of time. Their model consists of two equations representing the goods market equilibrium (11*) and the money market equilibrium (12*):

$$Y_t = \alpha_0 + \alpha_1 R_t + a_{yt}, \tag{11*}$$

$$R_t = \beta_0 + \beta_1 Y_t + \beta_2 P_t + a_{rt}, \tag{12*}$$

where the coefficients α_i and β_j are assumed known, and a_{it} $(i = y, r)$ are random variables. The central bank's loss function to be minimized is

$$L = E\sum_{t=1}^{T} (Y_t - \tilde{Y})^2 \tag{13*}$$

where T is the number of days in planning time-horizon, and $\tilde{Y}$ is the target value of Y_t. Here we shall slightly change the time structure of equation (11*) as

$$Y_{t+1} = \alpha_0 + \alpha_1 R_t + a_{yt+1} \tag{11**}$$

and substitute (12*) for R_t into (11*), obtaining

$$Y_{t+1} = (\alpha_0 + \alpha_1 \beta_0) + \alpha_1 \beta_1 Y_t + \alpha_1 \beta_2 P_t + (\alpha_1 a_{rt} + a_{yt+1}). \tag{14*}$$

Letting $\tilde{P}$ be the desired stock of reserves that equilibrates Y_t and Y_{t+1} with $\tilde{Y}$ in (14*) without random terms, i.e.,

$$\tilde{Y} = (\alpha_0 + \alpha_1 \beta_0) + \alpha_1 \beta_1 \tilde{Y} + \alpha_1 \beta_2 \tilde{P}. \tag{15*}$$

Subtracting (15*) from (14*) yields

$$(Y_{t+1} - \tilde{Y}) = \alpha_1 \beta_1 (Y_t - \tilde{Y}) + \alpha_1 \beta_2 (P_t - \tilde{P}) + (\alpha_1 a_{rt} + a_{yt+1}). \tag{16*}$$

Equation (16*) corresponds to (22b) with $x_1 = 0$. Thus the present problem will be able to be solved by applying Theorem 3 above.

5.3. Controllers for Linear Systems with Stochastic Coefficients

We now turn to the perfect state information case, where a linear discrete-time system contains uncertain parameters as well as additive disturbances. In particular, the system in question is of the following form:

$$x(t) = [A + Z_t]x(t-1) + [B + W_t]v(t) + \xi(t) \tag{41}$$

for $t = 1, \ldots, \beta$ (β: a terminal time period) with given initial state $x(0) = x_0$,

where

x	n vector of state variables,
v	m vector of control variables,
ξ	n vector of random disturbances,
A, B	constant coefficient matrices,
Z_t, W_t	stochastic coefficient matrices, of appropriate sizes.

We assume:

elements of $\xi(t)$, Z_t and W_t are all contemporaneously correlated with one another, (42a)

$$EZ_t = 0, \qquad EW_t = 0, \qquad E\xi(t) = 0, \tag{42b}$$

$$\operatorname{cov}(\xi(t), \xi(s)) = R\delta_{st}, \qquad (\delta\text{: Kronecker delta}). \tag{42c}$$

The cost function to be minimized is the conventional one (cf. (1))

$$J(x_0) \equiv E\left\{ x^T(\beta)\Gamma x(\beta) + \sum_{t=1}^{\beta} \left(x^T(t-1)\Xi x(t-1) + v^T(t)\Phi v(t) \right) \right\}, \tag{43}$$

where E represents an expectation, and Γ, Ξ, and Φ are assumed to be constant positive semidefinite matrices, with Φ positive definite.

Letting β be a given finite positive integer, we consider the finite time-horizon problem of minimizing J in (43) subject to the linear stochastic system (41) under assumptions (42).

Defining

$$\begin{aligned} g_t(x(t-1), v(t), \xi(t)) \equiv {} & ([A + Z_t]x(t-1) + [B + W_t]v(t) + \xi(t))^T \Xi \\ & \times ([A + Z_t]x(t-1) + [B + W_t]v(t) + \xi(t)) \\ & + v^T(t)\Phi v(t) \qquad \text{for} \quad t = 1, \ldots, \beta - 1, \end{aligned} \tag{44a}$$

$$\begin{aligned} & g_\beta(x(\beta-1), v(\beta), \xi(\beta)) \\ & \quad \equiv ([A + Z_\beta]x(\beta-1) + [B + W_\beta]v(\beta) + \xi(\beta))^T \Gamma \\ & \quad \times ([A + Z_\beta]x(\beta-1) + [B + W_\beta]v(\beta) + \xi(\beta)) + v^T(\beta)\Phi v(\beta) \end{aligned} \tag{44b}$$

we rewrite (43) as

$$J(x_0) = E\left\{ \sum_{t=1}^{\beta} g_t(x(t-1), v(t), \xi(t)) \right\} + x_0^T \Xi x_0 \tag{43'}$$

Let $J^*(x_0)$ be the optimal value of the expected cost $J(x_0)$ in (43′). Here, note that the probability measures characterizing $\xi(t)$, Z_t, and W_t do not depend on prior values of these variables and parameters. Also, control vector $v(t)$, to be optimal, is dependent on all the future optimization programs, in view of the optimality principle in dynamic programming. Thus $J^*(x_0)$, the optimal $J(x_0)$ in (43′), may be written as

$$J^*(x_0) = \min_{v(1)}\Big[\underset{t=1}{E}\Big\{g_1(x_0, v(1), \xi(1)) + \min_{v(2)}\Big[\underset{t=2}{E}\Big\{g_2(x(1), v(2), \xi(2)) + \cdots + \min_{v(\beta)}\Big[\underset{t=\beta}{E} g_\beta(x(\beta-1), v(\beta), \xi(\beta))\Big]\Big\}\Big]\cdots\Big\}\Big] + x_0^T \Xi x_0, \tag{45}$$

where the expectation E over $t = \tau$ is conditional on $x(\tau - 1)$ and $v(\tau)$, $\tau = 1, \ldots, \beta$.

Defining

$$J(\beta, \beta) \equiv \underset{t=\beta}{E} g_\beta(x(\beta-1), v(\beta), \xi(\beta)), \tag{46a}$$

$$J(\beta, \beta-1) \equiv \underset{t=\beta-1}{E}\Big\{g_{\beta-1}(x(\beta-2), v(\beta-1), \xi(\beta-1)) + \min_{v(\beta)} J(\beta, \beta)\Big\}, \tag{46b}$$

$$\vdots$$

$$J(\beta, \tau) \equiv \underset{t=\tau}{E}\Big\{g_\tau(x(\tau-1), v(\tau), \xi(\tau)) + \min_{v(\tau+1)} J(\beta, \tau+1)\Big\}, \tag{46c}$$

$$\vdots$$

$$J(\beta, 1) \equiv \underset{t=1}{E}\Big\{g_1(x_0, v(1), \xi(1)) + \min_{v(2)} J(\beta, 2)\Big\} + x_0^T \Xi x_0, \tag{46d}$$

we apply the optimality principle in dynamic programming. Thus, in the first place, the differential of $J(\beta, \beta)$ with respect to $v(\beta)$ is set equal to zero, i.e.,

$$\begin{aligned} 0 &= \frac{1}{2}\frac{\partial J(\beta, \beta)}{\partial v(\beta)} \\ &= B^T \Gamma B v(\beta) + B^T \Gamma A x(\beta-1) + \Phi v(\beta) + E\big(W_\beta^T \Gamma W_\beta\big) v(\beta) \\ &\quad + E\big(W_\beta^T \Gamma Z_\beta\big) x(\beta-1) + E\big(W_\beta^T \Gamma \xi(\beta)\big) \end{aligned} \tag{47}$$

since from (46a), (44b), and (43b)

$$\begin{aligned} J(\beta, \beta) &= (Ax(\beta-1) + Bv(\beta))^T \Gamma (Ax(\beta-1) + Bv(\beta)) \\ &\quad + v^T(\beta) \Phi v(\beta) + E\big\{(Z_\beta x(\beta-1) + W_\beta v(\beta) + \xi(\beta))^T \\ &\qquad \times \Gamma (Z_\beta x(\beta-1) + W_\beta v(\beta) + \xi(\beta))\big\}. \end{aligned} \tag{48a}$$

It follows from (47) that the optimal control at time β is determined by (cf. Chow (1975, p. 230) for a similar derivation)

$$v(\beta) = -K(\beta)x(\beta-1) - k(\beta) \tag{49a}$$

where

$$K(\beta) \equiv \left[\Phi + B^T\Gamma B + E\left(W_\beta^T \Gamma W_\beta\right)\right]^{-1}\left[B^T\Gamma A + E\left(W_\beta^T\Gamma Z_\beta\right)\right], \tag{50a}$$

$$k(\beta) \equiv \left[\Phi + B^T\Gamma B + E\left(W_\beta^T\Gamma W_\beta\right)\right]^{-1} E\left(W_\beta^T\Gamma\xi(\beta)\right). \tag{51a}$$

Substituting (49a) into (48a), we have the minimal value of $J(\beta, \beta)$ as

$$\begin{aligned} J^*(\beta, \beta) = x^T(\beta-1)&\left\{\left[A - BK(\beta)\right]^T\Gamma\left[A - BK(\beta)\right] + K^T(\beta)\Phi K(\beta)\right.\\ &\left.+ E\left[(Z_\beta - W_\beta K(\beta))^T\Gamma(Z_\beta - W_\beta K(\beta))\right]\right\}\\ &\times x(\beta-1) + k^T(\beta)\left[\Phi + B^T\Gamma B + E\left(W_\beta^T\Gamma W_\beta\right)\right]k(\beta)\\ &- 2k^T(\beta)E\left(W_\beta^T\Gamma\xi(\beta)\right) - 2k^T(\beta)\\ &\times\left\{B^T\Gamma\left[A - BK(\beta)\right] - \Phi K(\beta)\right.\\ &\left.+ E\left(W_\beta^T\Gamma\left[Z_\beta - W_\beta K(\beta)\right]\right)\right\}x(\beta-1)\\ &+ 2E\left(\xi^T(\beta)\Gamma(Z_\beta - W_\beta K(\beta))\right]x(\beta-1) + E\left(\xi^T(\beta)\Gamma\xi(\beta)\right) \end{aligned}$$

which is reduced to

$$\begin{aligned} J^*(\beta, \beta) = x^T(\beta-1)&\left\{\left[A - BK(\beta)\right]^T\Gamma\left[A - BK(\beta)\right] + K^T(\beta)\Phi K(\beta)\right.\\ &\left.+ E\left[(Z_\beta - W_\beta K(\beta))^T\Gamma(Z_\beta - W_\beta K(\beta))\right]\right\}\\ &\times x(\beta-1) + 2E\left[\xi^T(\beta)\Gamma(Z_\beta - W_\beta K(\beta))\right]x(\beta-1)\\ &- E\left(\xi^T(\beta)\Gamma W_\beta\right)k(\beta) + \operatorname{tr}(\Gamma R) \end{aligned} \tag{52a}$$

by virtue of (50a), (51a), and

$$E(z^T M y) = \operatorname{tr}(M\operatorname{cov}(y, z)) \tag{53}$$

which holds in the case that y or z has zero mean. (Cf. Lemma 1.)

Secondly, the differential of $J(\beta, \beta-1)$ with respect to $v(\beta-1)$ is set equal to zero. In view of (46b), (44a), and (52a), we get

$$\begin{aligned} J(\beta, \beta-1) = E&\left\{\left(\left[A + Z_{\beta-1}\right]x(\beta-2) + \left[B + W_{\beta-1}\right]v(\beta-1)\right.\right.\\ &\left.+ \xi(\beta-1)\right)^T S(\beta-1) + 2\left[E\left(\xi^T(\beta)\Gamma Z_\beta\right)\right.\\ &\left.\left.- E\left(\xi^T(\beta)\Gamma W_\beta\right)K(\beta)\right]\right\}\\ &\times\left(\left[A + Z_{\beta-1}\right]x(\beta-2) + \left[B + W_{\beta-1}\right]v(\beta-1) + \xi(\beta-1)\right)\\ &- E\left(\xi^T(\beta)\Gamma W_\beta\right)k(\beta) + \operatorname{tr}(\Gamma R) + v^T(\beta-1)\Phi v(\beta-1) \end{aligned}$$

which is rewritten as

$$\begin{aligned}J(\beta, \beta-1) &= \left[Ax(\beta-2) + Bv(\beta-1)\right]^T S(\beta-1)\\&\quad\times\left[Ax(\beta-2) + Bv(\beta-1)\right]\\&\quad+ E\left\{\left[Z_{\beta-1}x(\beta-2) + W_{\beta-1}v(\beta-1) + \xi(\beta-1)\right]^T S(\beta-1)\right.\\&\qquad\left.\times\left[Z_{\beta-1}x(\beta-2) + W_{\beta-1}v(\beta-1) + \xi(\beta-1)\right]\right\}\\&\quad+ 2\left[E(\xi^T(\beta)\Gamma Z_\beta) - E(\xi^T(\beta)\Gamma W_\beta)K(\beta)\right]\\&\quad\times\left[Ax(\beta-2) + Bv(\beta-1)\right] - E(\xi^T(\beta)\Gamma W_\beta)k(\beta)\\&\quad+ \operatorname{tr}(\Gamma R) + v^T(\beta-1)\Phi v(\beta-1),\end{aligned} \tag{54}$$

where

$$\begin{aligned}S(\beta-1) &\equiv \left[A - BK(\beta)\right]^T\Gamma\left[A - BK(\beta)\right] + K^T(\beta)\Phi K(\beta)\\&\quad+ E\left[(Z_\beta - W_\beta K(\beta))^T\Gamma(Z_\beta - W_\beta K(\beta))\right] + \Xi\\&= A^T\Gamma\left[A - BK(\beta)\right] + E\left[Z_\beta^T\Gamma(Z_\beta - W_\beta K(\beta))\right] + \Xi\end{aligned} \tag{55a}$$

in view of (50a). Thus we have

$$\begin{aligned}0 &= \frac{1}{2}\frac{\partial J(\beta, \beta-1)}{\partial v(\beta-1)}\\&= \left[B^T S(\beta-1)B + E\left(W_{\beta-1}^T S(\beta-1)W_{\beta-1}\right) + \Phi\right]v(\beta-1)\\&\quad+ \left[B^T S(\beta-1)A + E\left(W_{\beta-1}^T S(\beta-1)Z_{\beta-1}\right)\right]x(\beta-2)\\&\quad+ E\left(W_{\beta-1}^T S(\beta-1)\xi(\beta-1)\right)\\&\quad+ B^T\left[E\left(Z_\beta^T\Gamma\xi(\beta)\right) - K^T(\beta)E\left(W_\beta^T\Gamma\xi(\beta)\right)\right],\end{aligned}$$

from which follows the optimal control at time $\beta - 1$

$$v(\beta-1) = -K(\beta-1)x(\beta-2) - k(\beta-1) \tag{49b}$$

where

$$\begin{aligned}K(\beta-1) &\equiv \left[\Phi + B^T S(\beta-1)B + E\left(W_{\beta-1}^T S(\beta-1)W_{\beta-1}\right)\right]^{-1}\\&\quad\times\left[B^T S(\beta-1)A + E\left(W_{\beta-1}^T S(\beta-1)Z_{\beta-1}\right)\right],\end{aligned} \tag{50b}$$

$$\begin{aligned}k(\beta-1) &\equiv \left[\Phi + B^T S(\beta-1)B + E\left(W_{\beta-1}^T S(\beta-1)W_{\beta-1}\right)\right]^{-1}\\&\quad\times\left\{E\left(W_{\beta-1}^T S(\beta-1)\xi(\beta-1)\right)\right.\\&\qquad\left.+ B^T\left[E\left(Z_\beta^T\Gamma\xi(\beta)\right) - \left(A^T\Gamma B + E\left(Z_\beta^T\Gamma W_\beta\right)\right)k(\beta)\right]\right\}\end{aligned} \tag{51b}$$

in view of (50a) and (51a). (For simplicity, we write hereafter $E_t(W^T S(t)\xi)$ instead of $E(W_t^T S(t)\xi(t))$ and the like.)

Substituting (49b) into (54) yields the minimal value of $J(\beta, \beta-1)$ as

$$\begin{aligned}
&J^*(\beta, \beta-1)\\
&\quad = x^T(\beta-2)\Big\{\big[A - BK(\beta-1)\big]^T S(\beta-1)\big[A - BK(\beta-1)\big]\\
&\qquad + \underset{\beta-1}{E}\big[Z - WK(\beta-1)\big]^T S(\beta-1)\\
&\qquad \times \big[Z - WK(\beta-1)\big] + K^T(\beta-1)\Phi K(\beta-1)\Big\}\\
&\quad \times x(\beta-2) + 2\Big[\underset{\beta-1}{E}\big(\xi^T S(\beta-1)Z\big) + \underset{\beta}{E}(\xi^T\Gamma Z)A\\
&\qquad - \underset{\beta}{E}(\xi^T\Gamma W)K(\beta)A\Big]x(\beta-2) - 2k^T(\beta-1)\\
&\quad \times \Big[B^T S(\beta-1)A + \underset{\beta-1}{E}\big(W^T S(\beta-1)Z\big)\Big]x(\beta-2)\\
&\quad - k^T(\beta-1)\Big[\Phi + B^T S(\beta-1)B + \underset{\beta-1}{E}\big(W^T S(\beta-1)W\big)\Big]\\
&\quad \times k(\beta-1) - \underset{\beta}{E}(\xi^T\Gamma W)k(\beta) + \operatorname{tr}(\Gamma R) + \operatorname{tr}(S(\beta-1)R) \qquad (52b)
\end{aligned}$$

by virtue of (50a), (50b), (51a), and (51b).

Third, the differential of $J(\beta, \beta-2)$ with respect to $v(\beta-2)$ is equalized to zero. In view of (46c), (44a), and (52b), we have

$$\begin{aligned}
J(\beta, \beta-2) &= \big[Ax(\beta-3) + Bv(\beta-2)\big]^T\\
&\quad \times S(\beta-2)\big[Ax(\beta-3) + Bv(\beta-2)\big]\\
&\quad + E\big[Z_{\beta-2}x(\beta-3) + W_{\beta-2}v(\beta-2) + \xi(\beta-2)\big]^T S(\beta-2)\\
&\quad \times \big[Z_{\beta-2}x(\beta-3) + W_{\beta-2}v(\beta-2) + \xi(\beta-2)\big]\\
&\quad + 2\Big\{\underset{\beta-1}{E}\big(\xi^T S(\beta-1)Z\big) + \underset{\beta}{E}(\xi^T\Gamma Z)A - \underset{\beta}{E}(\xi^T\Gamma W)K(\beta)A\\
&\qquad - k^T(\beta-1)\Big[B^T S(\beta-1)A + \underset{\beta-1}{E}\big(W^T S(\beta-1)Z\big)\Big]\Big\}\\
&\quad \times \big[Ax(\beta-3) + Bv(\beta-2)\big] + v^T(\beta-2)\Phi v(\beta-2)\\
&\quad - \underset{\beta}{E}(\xi^T\Gamma W)k(\beta) + \operatorname{tr}(\Gamma R) + \operatorname{tr}(S(\beta-1)R) - k^T(\beta-1)\\
&\quad \times \Big[\Phi + B^T S(\beta-1)B + \underset{\beta-1}{E}\big(W^T S(\beta-1)W\big)\Big]k(\beta-1)
\end{aligned}$$

where

$$
\begin{aligned}
S(\beta-2) \equiv & \left[A - BK(\beta-1)\right]^T S(\beta-1)\left[A - BK(\beta-1)\right] \\
& + K^T(\beta-1)\Phi K(\beta-1) \\
& + \underset{\beta-1}{E}\left[(Z - WK(\beta-1))^T S(\beta-1)(Z - WK(\beta-1))\right] + \Xi \\
= & A^T S(\beta-1)\left[A - BK(\beta)\right] \\
& + \underset{\beta-1}{E}\left[Z^T S(\beta-1)(Z - WK(\beta-1))\right] + \Xi
\end{aligned} \tag{55b}
$$

in view of (50b). Thus we have

$$
\begin{aligned}
0 = & \frac{1}{2}\frac{\partial J(\beta, \beta-2)}{\partial v(\beta-2)} \\
= & \left[B^T S(\beta-2)B + \underset{\beta-2}{E}\left(W^T S(\beta-2)W\right) + \Phi\right]v(\beta-2) \\
& + \left[B^T S(\beta-2)A + \underset{\beta-2}{E}\left(W^T S(\beta-2)Z\right)\right]x(\beta-3) \\
& + \underset{\beta-2}{E}\left(W^T S(\beta-2)\xi\right) + B^T\Big[\underset{\beta-1}{E}\left(Z^T S(\beta-1)\xi\right) - \Big(A^T S(\beta-1)B \\
& \qquad + \underset{\beta-1}{E}\left(Z^T S(\beta-1)W\right)\Big)k(\beta-1)\Big] \\
& + B^T A^T\left[\underset{\beta}{E}(Z^T\Gamma\xi) - \left(A^T\Gamma B + \underset{\beta}{E}(Z^T\Gamma W)\right)k(\beta)\right]
\end{aligned}
$$

from which we get the optimal control at time $\beta - 2$ as

$$
v(\beta-2) = -K(\beta-2)x(\beta-3) - k(\beta-2), \tag{49c}
$$

where

$$
\begin{aligned}
K(\beta-2) \equiv & \left[\Phi + B^T S(\beta-2)B + \underset{\beta-2}{E}\left(W^T S(\beta-2)W\right)\right]^{-1} \\
& \times\left[B^T S(\beta-2)A + \underset{\beta-2}{E}\left(W^T S(\beta-2)Z\right)\right]
\end{aligned} \tag{50c}
$$

$$
\begin{aligned}
k(\beta-3) \equiv & \left[\Phi + B^T S(\beta-2)B + \underset{\beta-2}{E}\left(W^T S(\beta-2)W\right)\right]^{-1} \\
& \times\Big\{\underset{\beta-2}{E}\left(W^T S(\beta-2)\xi\right) + B^T\Big[\underset{\beta-1}{E}\left(Z^T S(\beta-1)\xi\right) \\
& \qquad - \left(A^T S(\beta-1)B + \underset{\beta-1}{E}\left(Z^T S(\beta-1)W\right)k(\beta-1)\right)\Big] \\
& + B^T A^T\left[\underset{\beta}{E}(Z^T\Gamma\xi) - \left(A^T\Gamma B + \underset{\beta}{E}(Z^T\Gamma W)\right)k(\beta)\right]\Big\}.
\end{aligned} \tag{51c}
$$

Carrying out these calculations consecutively, we obtain the following general rule of optimal control.

Theorem 4. *The optimal feedback control to the finite time-horizon problem of minimizing J in* (43), *subject to the linear system* (41) *under the assumptions* (42), *is given by*

$$v(t) = -K(t)x(t-1) - k(t) \qquad \text{for} \quad t = 1, 2, \ldots, \beta \tag{49}$$

where

$$K(t) \equiv \left[\Phi + B^T S(t) B + E\left(W_t^T S(t) W_t\right)\right]^{-1}\left[B^T S(t) A + E\left(W_t^T S(t) Z_t\right)\right] \tag{50}$$

$$\begin{aligned} k(t) \equiv \left[\Phi + B^T S(t) B + E\left(W_t^T S(t) W_t\right)\right]^{-1} & \\ \times \Big\{ E\left(W_t^T S(t)\xi(t)\right) + B^T\Big[E\left(Z_{t+1}^T S(t+1)\xi(t+1)\right) & \\ - \left(A^T S(t+1) B + E\left(Z_{t+1}^T S(t+1) W_{t+1}\right)\right) & \\ \times k(t+1)\Big] & \\ + B^T A^T\Big[E\left(Z_{t+2}^T S(t+2)\xi(t+2)\right) & \\ - \left(A^T S(t+2) B + E\left(Z_{t+2}^T S(t+2) W_{t+2}\right)\right) k(t+2)\Big] & \\ + \ldots + B^T (A^T)^{\beta - t - 1}\Big[E\left(Z_\beta^T S(\beta)\xi(\beta)\right) & \\ - \left(A^T S(\beta) B + E\left(Z_\beta^T S(\beta) W_\beta\right)\right) & \\ \times k(\beta)\Big]\Big\}, & \end{aligned} \tag{51}$$

$S(t)$ is calculated backward in time, beginning from

$$S(\beta) = \Gamma,$$

by formula (55) *or* (55′) *below for $t = \beta - 1, \beta - 2, \ldots, 2, 1$, one after another*:

$$\begin{aligned} S(t) = A^T S(t+1)\left[A - BK(t+1)\right] + E\left(Z_{t+1}^T S(t+1) Z_{t+1}\right) + \Xi & \\ - E\left(Z_{t+1}^T S(t+1) W_{t+1}\right) K(t+1) & \end{aligned} \tag{55}$$

or, equivalently,

$$\begin{aligned} S(t) = \left[A - BK(t+1)\right]^T S(t+1)\left[A - BK(t+1)\right] & \\ + K^T(t+1)\Phi K(t+1) + \Xi & \\ + E\left\{\left[Z_{t+1} - W_{t+1}K(t+1)\right]^T S(t+1)\left[Z_{t+1} - W_{t+1}K(t+1)\right]\right\} & \end{aligned} \tag{55′}$$

Corollary. *If assumption* (42*a*) *is replaced by*

$\xi(t), Z_t$ *and* W_t *are contemporaneously uncorrelated with one another,* $$(42^\dagger a)$$

then the above optimal control reduces to

$$v(t) = -K^*(t)x(t-1) \qquad \text{for} \quad t = 1, 2, \ldots, \beta \tag{49†}$$

where

$$K^*(t) \equiv \left[\Phi + B^T S^*(t) B + E\left(W_t^T S^*(t) W_t\right)\right]^{-1} \left(B^T S^*(t) A\right), \tag{50†}$$

$$S^*(t) = A^T S^*(t+1)\left[A - BK^*(t+1)\right] + E\left(Z_{t+1}^T S^*(t+1) Z_{t+1}\right) + \Xi \tag{55†}$$

for $t = \beta - 1, \ldots, 2, 1$ *with* $S^*(\beta) = \Gamma$.

Next, we consider the infinite time-horizon problem of minimizing the expected value of cost function

$$\tilde{J}(x_0) = E \sum_{t=1}^{\infty} \left(x^T(t-1)\Xi x(t-1) + v^T(t)\Phi v(t)\right) \tag{56}$$

for given $x(0) = x_0$, with respect to $v(t)$ $(t = 1, 2, \ldots)$, subject to system (41) under the assumptions (42). (Notations are the same as before.)

We assume that all random variables have converged to their limiting distributions. In such a stationary state, the optimal control is deduced from Theorem 4 as follows.

Theorem 5. *Under stationary conditions, the optimal feedback control for the infinite time-horizon problem of minimizing* $\tilde{J}$ *in* (56) *subject to the linear stochastic system* (41) *under the assumptions* (42) *is given by*

$$v(t) = -Kx(t-1) - k \qquad \text{for} \quad t = 1, 2, \ldots \tag{57}$$

where

$$K \equiv \left[\Phi + B^T S B + E(W^T S W)\right]^{-1}\left[B^T S A + E(W^T S Z)\right] \tag{58}$$

$$k \equiv \left[\Phi + B^T S B + E(W^T S W)\right]^{-1}\left\{E(W^T S \xi) + B^T p\right\}. \tag{59}$$

S and p are solutions to the following equations, respectively:

$$S = A^T S\left[A - BK\right] + E(Z^T S Z) - E(Z^T S W)K + \Xi \tag{60}$$

or, equivalently,

$$S = \left[A - BK\right]^T S\left[A - BK\right] + K^T \Phi K + \Xi + E\left\{\left[Z - WK\right]^T S\left[Z - WK\right]\right\}; \tag{60$'$}$$

$$p = \left[A - BK\right]^T p + E(Z^T S \xi) - K^T E(W^T S \xi). \tag{61}$$

PROOF. The result is obtainable by assuming that $K(t)$, $k(t)$, and $S(t)$ have converged to their limits K, k, and S, respectively. In particular, deriving (59) and (61) requires the convergence

$$\sum_{t=0}^{\infty} A^{t} = [I - A]^{-1}. \tag{62}$$

Then the convergence limit of (51) can be expressed as the combination of (59) with

$$\begin{aligned} p &= [I - A^T]^{-1}\{E(Z^T S\xi) - (A^T SB + E(Z^T SW))k\} \\ &= [I - A^T]^{-1}\{E(Z^T S\xi) - K^T(E(W^T S\xi) + B^T p)\}, \end{aligned}$$

from which (61) follows. □

REMARK (Turnovsky's Lagrange Multiplier Method). The result of Theorem 5 above is alternatively derived by applying to assumed asymptotic relationships the Lagrange multiplier method which was adopted by Turnovsky (1976). We analyze this method in detail in order to understand its derivational process.

Turnovsky chooses $v(t)$ so as to minimize the asymptotic cost

$$E\{x^T(t)\Xi x(t) + v^T(t)\Phi v(t)\}, \tag{63}$$

where Ξ and Φ are positive definite matrices, subject to our dynamic equation (41) under all the conditions (42) held. He begins with the assumption of a feedback control law of the form

$$v(t) = -Kx(t-1) - k, \tag{64}$$

where K and k remain to be determined. Substituting (64) into (41) yields

$$x(t) = [A - BK + Z_t - W_t K]x(t-1) - [B + W_t]k + \xi(t) \tag{65}$$

from which we determine that the asymptotic expectation of $x(t)$, denoted by μ, must satisfy

$$\mu = [A - BK]\mu - Bk. \tag{66}$$

Define the asymptotic expectations $E(x(t)x^T(t)) = X$ and $E(\xi(t)\xi^T(t)) = R$. Then (65) implies the following asymptotic relationship (note that $E(y_t w_{t-1}) = 0$ for any random variables y and w):

$$\begin{aligned} X = {} & [A - BK]X[A - BK]^T + E\{[Z - WK]X[Z - WK]^T\} \\ & + Bkk^T B^T + E(Wkk^T W^T) + R - 2[A - BK]\mu k^T B^T \\ & - 2E\{[Z - WK]\mu k^T W^T\} \\ & + 2E\{[Z - WK]\mu\xi^T\} - 2E(Wk\xi^T). \end{aligned} \tag{67}$$

On the other hand, by applying Lemma 1, the cost expression (63) can be rewritten as

$$\left(Ex^T(t)\right)\Xi\left(Ex(t)\right)+\left(Ev^T(t)\right)\Phi\left(Ev(t)\right)+\operatorname{tr}\left[\Xi\operatorname{cov}(x(t))+\Phi\operatorname{cov}(v(t))\right] \tag{63'}$$

where $\operatorname{cov}(x(t))$ and $\operatorname{cov}(v(t))$ denote the variance-covariance matrices of $x(t)$ and $v(t)$, respectively. Taking (64) into account, we get

$$\begin{aligned}\operatorname{cov}(v(t)) &= E\left\{(v(t)-Ev(t))(v(t)-Ev(t))^T\right\}\\ &= KE\left\{(x(t-1)-\mu)(x(t-1)-\mu)^T\right\}K^T\\ &= K\operatorname{cov}(x(t-1))K^T.\end{aligned}$$

Since asymptotically $\operatorname{cov}(x(t))=\operatorname{cov}(x(t-1))=X-\mu\mu^T$, we have

$$\operatorname{cov}(v(t))=KXK^T-K\mu\mu^TK^T.$$

Substituting all these asymptotic relations into (63′) yields

$$\mu^T\Xi\mu+(K\mu+k)^T\Phi(K\mu+k)+\operatorname{tr}\left[\Xi(X-\mu\mu^T)+\Phi(KXK^T-K\mu\mu^TK^T)\right]$$

which is reduced to

$$2k^T\Phi K\mu+k^T\Phi k+\operatorname{tr}\left[(\Xi+K^T\Phi K)X\right] \tag{68}$$

since $\operatorname{tr}(\Xi\mu\mu^T)=\mu^T\Xi\mu$ and $\operatorname{tr}(\Phi K\mu\mu^TK^T)=\mu^TK^T\Phi K\mu$. Equation (68) is equivalent to

$$\operatorname{tr}\left[(\Xi+K^T\Phi K)X+2\Phi K\mu k^T+\Phi kk^T\right]. \tag{68'}$$

Thus our present problem reduces to minimizing (68′) subject to (66) and (67). Express the corresponding Lagrangian form (cf. Murata (1977, p. 262)) as follows:

$$\begin{aligned}L=&\operatorname{tr}\left[(\Xi+K^T\Phi K)X+2\Phi K\mu k^T+\Phi kk^T\right]\\ &+\operatorname{tr}\Big[S\Big\{(A-BK)X(A-BK)^T+R-X\\ &\qquad+E\left[(Z-WK)X(Z-WK)^T\right]+Bkk^TB^T+E(Wkk^TW^T)\\ &\qquad-2(A-BK)\mu k^TB^T-2E\left[(Z-WK)\mu k^TW^T\right]\\ &\qquad+2E\left[(Z-WK)\mu\xi^T\right]-2E(Wk\xi^T)\Big\}\Big]\\ &+2p^T\left[(A-BK)\mu-Bk-\mu\right],\end{aligned} \tag{69}$$

where S is the matrix of Lagrange multipliers associated with (67) and $2p$ is a column vector of Lagrange multipliers associated with (66). We are required to choose X, K, k, μ to minimize L in (69). Let us assume that S is symmetric, differentiate (69) with respect to X, K, μ, k, respectively, and set the results equal to zero (see the Appendix):

$$\begin{aligned}0 &= \frac{\partial L}{\partial X}\\ &= (\Xi + K^T\Phi K)^T + (A - BK)^T S(A - BK) - S\\ &\quad + E\left[(Z - WK)^T S(Z - WK)\right]\end{aligned} \tag{70a}$$

$$\begin{aligned}0 &= \frac{1}{2}\frac{\partial L}{\partial K}\\ &= \Phi KX + \Phi k\mu^T - B^TSAX + B^TSBKX + B^TSBk\mu^T - B^Tp\mu^T\\ &\quad + E(W^TSWKX - W^TSZX + W^TSWk\mu^T - W^TS\xi\mu^T)\\ &= \left[\Phi + B^TSB + E(W^TSW)\right]KX - \left[B^TSA + E(W^TSZ)\right]X\\ &\quad + \left[\Phi + B^TSB + E(W^TSW)\right]k\mu^T - \left[B^Tp + E(W^TS\xi)\right]\mu^T\end{aligned} \tag{70b}$$

$$\begin{aligned}0 &= \frac{1}{2}\frac{\partial L}{\partial \mu}\\ &= K^T\Phi k - (A - BK)^T SBk - E\left[(Z - WK)^T SWk - (Z - WK)^T S\xi\right]\\ &\quad + (A - BK - I)^T p\\ &= K^T\left[\Phi + B^TSB + E(W^TSW)\right]k - \left[A^TSB + E(ZSW)\right]k\\ &\quad + E\left[(Z - WK)^T S\xi\right] + (A - BK - I)^T p\end{aligned} \tag{70c}$$

$$\begin{aligned}0 &= \frac{1}{2}\frac{\partial L}{\partial k}\\ &= \Phi K\mu + \Phi k + B^TSBk + E(W^TSWk) - B^TS(A - BK)\mu\\ &\quad - E\left(W^TS(Z - WK)\mu\right) - E(W^TS\xi) - B^Tp\\ &= \left[\Phi + B^TSB + E(W^TSW)\right](K\mu + k) - \left[B^TSA + E(W^TSZ)\right]\mu\\ &\quad - E(W^TS\xi) - B^Tp.\end{aligned} \tag{70d}$$

Subtracting (70d) $\cdot\, \mu^T$ from (70b) yields

$$0 = \left\{\left[\Phi + B^TSB + E(W^TSW)\right]K - \left[B^TSA + E(W^TSZ)\right]\right\}(X - \mu\mu^T).$$

Hence, on the assumption that $X - \mu\mu^T$ is nonsingular, we get

$$K = \left[\Phi + B^TSB + E(W^TSW)\right]^{-1}\left[B^TSA + E(W^TSZ)\right]. \tag{58}$$

The substitution of (58) for K into (70c) yields

$$E\left[(Z - WK)^T S\xi\right] + (A - BK)^T p = p$$

which is nothing but (61) and gives the solution for p. Then, substituting (58) for K into (70d) yields the solution for k in form (59). Finally, we note that (70a) is the same as (60′). This completes the proof of Theorem 5 by the Turnovsky's Lagrange multiplier method.

For further developments parallel to our discussions on this topic, refer to Turnovsky (1977). The above Lagrange multiplier method will be utilized to derive Theorem 8 in Section 5.5.

5.4. Certainty Equivalence in Stochastic Systems Control by Theil

We mentioned the *certainty equivalence* principle in the remark on Theorem 1 in Section 5.1, where a relevant stochastic system contains an additive disturbance having zero mean. This principle is also applicable to the imperfect state information case as seen in Theorem 3 in Section 5.2. In both cases, the objective cost function (1) is quadratic in state variables and control variables separately and does not contain either linear or bilinear term in these variables. However, in his original argument for the principle, Theil (1957) assumed an objective function of the general quadratic form involving linear and bilinear terms. We examine here the principle in this respect and clarify the relationship between his optimal rule and ours.

We begin with the economic problem of Simon (1956), which is to determine at the beginning of each month the production level $P(t)$ of a commodity during that month. The forecasts of sales $S(t)$, $t = 1, \ldots, T$, are available subject to random errors, whose joint probability distribution is known. The initial inventory $I(0)$ and production $P(0)$ are given at some fixed levels, but the terminal inventory and production levels, $I(T)$ and $P(T)$, are not fixed. The definition of the inventory is

$$I(t) = I(t-1) + P(t) - S(t), \qquad t = 1, \ldots, T. \tag{17*}$$

The costs in each period consist of three kinds: 1) the costs of holding inventory and of run-outs resulting from inadequate inventory; 2) the costs associated with the production level; and 3) the costs associated with the change in level of production. Each of these three components is assumed to be quadratic in the cost function in question. Thus the expected value of total cost over the T periods we wish to minimize is expressed as

$$J = \underset{S(t)}{E} \sum_{t=1}^{T} \left\{ \alpha_1 \left[I(t) - I_c \right]^2 + \alpha_2 \left[P(t) - P_c \right]^2 + \alpha_3 \left[P(t) - P(t-1) \right]^2 \right\} \tag{18*}$$

where α_1, α_2, α_3, I_c, and P_c are known constants. Simon (1956) presupposed $P(t)$ to be a function ϕ_t of $S(1), \ldots, S(t-1)$; and optimally determined by the policy are $P(1)$ and expected values of ϕ_t $(t = 2, 3, \ldots, T)$.

In this example, $I(t)$ is a state variable, $P(t)$ is a control variable, and $S(t)$ is a stochastic exogenous variable which may be divided into two: mean value $\bar{S}(t)$ and the residual $\epsilon(t) \equiv \bar{S}(t) - S(t)$. Then (17*) becomes a specific representation of the state-space form (2′) in the corollary to Theorem 1 of Section 5.1: that is,

$$I(t) = I(t-1) + P(t) - \bar{S}(t) + \epsilon(t), \qquad t = 1, \ldots, T \tag{17**}$$

with $E\epsilon(t) = 0$ and $\operatorname{cov}(\epsilon(t), \epsilon(s)) = r\delta_{st}$ (δ_{st}: Kronecker delta). On the other hand, the objective function (18*) contains linear and bilinear terms as well as quadratic terms in state and control variables. Hence it differs from our conventional quadratic cost function such as (1) in Section 5.1. Function (18*) can be rewritten as

$$J = E\left\{\alpha_1 \tilde{x}^T \tilde{x} + a_1 \tilde{x} + \tilde{v}^T \Phi \tilde{v} + a_2 \tilde{v}\right\} + \text{constant}, \tag{19*}$$

where superscript T denotes transposition,

$$\tilde{x} \equiv \begin{bmatrix} I(1) - I(0) \\ I(2) - I(0) \\ \vdots \\ I(T) - I(0) \end{bmatrix}, \qquad \tilde{v} \equiv \begin{bmatrix} P(1) \\ P(2) \\ \vdots \\ P(T) \end{bmatrix}, \tag{20*}$$

$$a_1 \equiv -2\alpha_1(I_c - I(0))(1, \ldots, 1), \tag{21*a}$$

$$a_2 \equiv -2((\alpha_2 P_c + \alpha_3 P(0)), \alpha_2 P_c, \alpha_2 P_c, \ldots, \alpha_2 P_c), \tag{21*b}$$

$$\Phi \equiv \begin{bmatrix} \alpha_2 + 2\alpha_3 & -\alpha_3 & 0 & \cdots & 0 & 0 \\ -\alpha_3 & \alpha_2 + 2\alpha_3 & -\alpha_3 & \ddots & \vdots & \vdots \\ 0 & -\alpha_3 & \ddots & \ddots & 0 & 0 \\ 0 & 0 & \ddots & \ddots & -\alpha_3 & 0 \\ \vdots & \vdots & \ddots & \ddots & \alpha_2 + 2\alpha_3 & -\alpha_3 \\ 0 & 0 & \cdots & 0 & -\alpha_3 & \alpha_2 + \alpha_3 \end{bmatrix} \tag{21*c}$$

Employing the vector notations in (20*), the system of equations (17*), after iterative substitutions for $I(t-1)$, can be expressed as

$$\tilde{x} = \tilde{R}\tilde{v} - \tilde{R}\tilde{S}, \tag{22*}$$

where

$$\tilde{R} \equiv \begin{bmatrix} 1 & 0 & \cdots & \cdots & 0 \\ & & \ddots & & \vdots \\ 1 & 1 & & \ddots & \vdots \\ \vdots & \vdots & \ddots & & 0 \\ \vdots & \vdots & & \ddots & \\ 1 & 1 & \cdots & \cdots & 1 \end{bmatrix}, \qquad \tilde{S} \equiv \begin{bmatrix} S(1) \\ S(2) \\ \vdots \\ \vdots \\ S(T) \end{bmatrix}. \tag{23*}$$

Theil (1957) generalized the above problem by Simon as follows. To minimize the objective function

$$J = E\left\{ax + bv + \tfrac{1}{2}\left(x^T\tilde{\Xi}x + v^T\tilde{\Phi}v + x^T\Omega v + v^T\Omega^T x\right)\right\}, \tag{71}$$

with respect to v, subject to the system

$$x = Rv + s, \tag{72}$$

where

$$x \equiv \begin{bmatrix} x(1) \\ \vdots \\ x(T) \end{bmatrix}, \qquad x(t) \equiv \begin{bmatrix} x_1(t) \\ \vdots \\ x_n(t) \end{bmatrix}, \qquad v \equiv \begin{bmatrix} v(1) \\ \vdots \\ v(T) \end{bmatrix}, \qquad v(t) \equiv \begin{bmatrix} v_1(t) \\ \vdots \\ v_m(t) \end{bmatrix}, \tag{73}$$

a and b are constant row vectors, $\tilde{\Xi}$, $\tilde{\Phi}$, and Ω are constant matrices of appropriate sizes ($\tilde{\Xi}$ and $\tilde{\Phi}$ symmetric positive definite), R is an $nT \times mT$ matrix of constant elements that can be partitioned as

$$R = \begin{bmatrix} R_{11} & 0 & \cdots & \cdots & \cdots & 0 \\ & & \ddots & & & \vdots \\ R_{21} & R_{22} & & \ddots & & \vdots \\ \vdots & \vdots & \ddots & & \ddots & \vdots \\ \vdots & \vdots & & \ddots & & 0 \\ \vdots & \vdots & & & \ddots & \\ R_{T1} & R_{T2} & \cdots & \cdots & \cdots & R_{TT} \end{bmatrix}, \tag{74}$$

each submatrix of which has dimension $n \times m$, and s is a column vector of

nT random elements whose joint distribution is independent of v:

$$s \equiv \begin{bmatrix} s(1) \\ \vdots \\ s(T) \end{bmatrix}, \qquad s(t) \equiv \begin{bmatrix} s_1(t) \\ \vdots \\ s_n(t) \end{bmatrix}. \tag{75}$$

Note that system (72) is general enough to include the conventional state-space form of equations such as

$$x(t) = Ax(t-1) + Bv(t) + \xi(t), \qquad t = 1, \ldots, T. \tag{76}$$

Because iterative substitutions in (76) for $x(t-1)$ yield

$$\begin{aligned} x(1) &= Ax(0) + Bv(1) + \xi(1), \\ x(2) &= A^2x(0) + ABv(1) + Bv(2) + (A\xi(1) + \xi(2)), \\ x(3) &= A^3x(0) + A^2Bv(1) + ABv(2) + Bv(3) \\ &\quad + \big(A^2\xi(1) + A\xi(2) + \xi(3)\big), \\ &\vdots \\ x(T) &= A^Tx(0) + A^{T-1}Bv(1) + \cdots + ABv(T-1) + Bv(T) \\ &\quad + \sum_{t=1}^{T} A^{t-1}\xi(T-t+1). \end{aligned} \tag{76$'$}$$

Then, putting

$$s(\tau) = \sum_{t=1}^{\tau} A^{t-1}\xi(\tau - t + 1) + A^{\tau}x(0), \qquad \tau = 1, \ldots, T \tag{77}$$

and

$$\begin{aligned} &R_{11} = B, \\ &R_{21} = AB, R_{22} = B \\ &\vdots \\ &R_{T1} = A^{T-1}B, R_{T2} = A^{T-2}B, \ldots, R_{TT} = B, \end{aligned} \tag{78}$$

we reach system (72).

Following Simon's line of approach, we now establish the certainty equivalence principle for the stochastic control problem generalized above. (Refer to Theil (1961, Section 8.6) as well as Theil (1957) for the proof of Theorem 6 below.)

Theorem 6 (First-Period Certainty Equivalence). *Consider the problem of minimizing the cost function value J in (71) with respect to instruments v subject to the constraint (72), where s is a vector of random elements with* *mean value $\bar{s}$ and a finite covariance matrix, its distribution being the same for whatever v. Suppose that an optimal strategy for this problem exists and that it can be written as*

$$v^* = \begin{bmatrix} v^*(1) \\ v^*(2) = \phi_2(s(1)) \\ v^*(3) = \phi_3(s(1), s(2)) \\ \vdots \\ v^*(T) = \phi_T(s(1), \ldots, S(T-1)) \end{bmatrix}. \tag{79}$$

Then the strategy v^* implies the same first-period decision $v^*(1)$ as the control vector v, denoted $\bar{v}$, that minimizes $\bar{J}$ defined as

$$\bar{J} = ax + bv + \tfrac{1}{2}(x^T\tilde{\Xi}x + v^T\tilde{\Phi}v + x^T\Omega v + v^T\Omega^T x) \tag{71'}$$

under the deterministic system

$$x = Rv + \bar{s}, \tag{72'}$$

which is nothing but equation (72) *with s replaced by $\bar{s}$. Finally, the optimal strategy v^* is unique.*

PROOF. Substituting (72) for x into (71) yields

$$J = E\big(h_0 + h_1 v + \tfrac{1}{2}v^T H v\big), \tag{80}$$

where

$$h_0 \equiv as + \tfrac{1}{2}s^T\tilde{\Xi}s, \tag{81a}$$

$$h_1 \equiv b + aR + s^T(\tilde{\Xi}R + \Omega), \qquad (1 \times mT) \tag{81b}$$

$$H \equiv R^T\tilde{\Xi}R + \tilde{\Phi} + R^T\Omega + \Omega^T R, \qquad (mT \times mT). \tag{81c}$$

Note that H is a symmetric positive definite and nonstochastic matrix and that

$$Eh_0 = a\bar{s} + \tfrac{1}{2}\big(\bar{s}^T\tilde{\Xi}\bar{s} + \operatorname{tr}\big(\tilde{\Xi}\operatorname{cov}(s,s)\big)\big) \tag{81'a}$$

in view of Lemma 1 in Section 5.1, whence Eh_0 becomes independent of v. Any strategy v deviating from v^* can be expressed as

$$v = v^* + \lambda w, \tag{82}$$

where λ is a scalar and w is a vector of the type

$$w \equiv \begin{bmatrix} w_1 \\ w_2(s(1)) \\ \vdots \\ w_T(s(1), \ldots, s(T-1)) \end{bmatrix}. \tag{83}$$

Substitution of (82) into (80) gives

$$J = Eh_0 + E(h_1 v^*) + \tfrac{1}{2}E(v^{*T}Hv^*)$$
$$+ \lambda\left[E(h_1 w) + E(w^T Hv^*)\right] + \tfrac{1}{2}\lambda^2 E(w^T Hw). \tag{80'}$$

The second and third terms on the right-hand side of (80′) are independent of λ and w, and hence of v. Thus they are disregarded in the minimization process. The first-order condition for J to take on a minimum value at $\lambda = 0$ is that the differential of J with respect to λ evaluated at $\lambda = 0$ vanish, i.e.,

$$E(h_1 w) + E(w^T Hv^*) = 0 \qquad \text{for any } w. \tag{84}$$

A sufficient condition for the extremum point to be indeed a minimum point is that the second-order derivative $E(w^T Hw)$ of J with respect to λ be positive definite, which follows from the fact that matrix H is positive definite. Consider then (84) and specify w in the following mT alternative ways. First, choose w such that its first element equals 1 and all others zeros; secondly, such that its second element equals 1 and all others zeros; and so on. The result brings forth a system of mT equations:

$$Eh_1^T + HEv^* = 0, \tag{85}$$

from which we obtain

$$Ev^* = -H^{-1}\bar{h}_1^T, \tag{85'}$$

where

$$\bar{h}_1 = b + aR + \bar{s}^T(\tilde{\Xi}R + \Omega). \tag{81'b}$$

Next, we consider the certainty case. Substituting (72′) into (71′) yields

$$\bar{J} = \bar{h}_0 + \bar{h}_1 v + \tfrac{1}{2}v^T Hv, \tag{86}$$

where $\bar{h}_1$ and H are those of (81′b) and (81c), respectively, and

$$\bar{h}_0 \equiv a\bar{s} + \tfrac{1}{2}\bar{s}^T\tilde{\Xi}\bar{s}. \tag{81''a}$$

The v that minimizes $\bar{J}$ of (86) is derived by setting the differential of $\bar{J}$ with respect to v equal to zero, i.e.,

$$0 = \frac{d\bar{J}}{dv} = \bar{h}_1^T + Hv. \tag{87}$$

Thus the optimal strategy $\bar{v}$ is obtained as

$$\bar{v} = -H^{-1}\bar{h}_1^T. \tag{87'}$$

The first subvector $v^*(1)$ of v^* is nonstochastic (cf. (79)), and hence it must be equal to the corresponding subvector of $\bar{v}$ in view of the fact that $Ev^* = \bar{v}$.

Finally, the uniqueness of the optimal strategy v^* is guaranteed by the fact that matrix H of (81c) is positive definite, because then J of (80) is strictly convex in v. (Cf. Corollary to Theorem 11 in Murata (1977, Section 7.2).) □

REMARK 1. By virtue of Theorem 6 above, the policymaker in the present stochastic situation can decide his optimal first-period strategy $v(1)$ by replacing random term s by its expectation $\bar{s}$. At the beginning of the second period the same situation arises, so he should apply the same procedure, with $x(1)$, $v(1)$, R_{t1} ($t = 1, \ldots, T$) deleted, and replace $s(2)$, $s(3), \ldots$ by their conditional means, given the information available at the end of $t = 1$.

REMARK 2. In Theorem 6, we assume that the distribution of random vector s is the same for any strategy v. This assumption was restated by Theil (1964, p. 130) as "the distribution of the subvector $s(t)$ is independent of $v(t')$ for $t, t' = 1, \ldots, T$ and $t \geqslant t'$." However, as Duchan (1974) noticed, the "restatement" is stronger than the former assumption of ours. "To see this," continued Duchan, "consider a two-period problem in which the only information available at the end of the first period is the realization of $s(1)$ so that a strategy for $v(2)$ is just some function of $s(1)$. When $s(1)$ and $s(2)$ are dependent, there is no particular reason to believe that the certainty equivalence strategy will be such that $v(2)$ and $s(2)$ are independent." Based on our assumption mentioned above, he gave a new proof of the first-period certainty equivalence strategy together with a new optimal strategy formula, by proceeding backward starting from the final period.

Now we want to clarify explicitly the relationship between the linear feedback control rule established in Theorem 1 (Section 5.1) and the first-period and other period certainty equivalence. In order to reduce Theil's formulation (71), (72) to the conventional optimal control scheme (1), (2), we set all elements of vectors a, b and matrix Ω at zeros and specify $\tilde{\Xi}$ and $\tilde{\Phi}$ as block-diagonal matrices:

$$\tilde{\Xi} \equiv \begin{bmatrix} \Xi & & & \\ & \ddots & & \\ & & \Xi & \\ & & & \Gamma \end{bmatrix}, \qquad \tilde{\Phi} \equiv \begin{bmatrix} \Phi & & & \\ & \Phi & & \\ & & \ddots & \\ & & & \Phi \end{bmatrix}. \tag{88}$$

Then, recalling (77) and (78) and knowing $\bar{s}(\tau) = A^{\tau}x(0)$, we can write out vector $\bar{h}_1$ and matrix H as below (instead of the notation T for the final period, we denote β in the following to avoid confusion):

$$\begin{aligned} \bar{h}_1 &= \bar{s}^T \tilde{\Xi} R \\ &= x^T(0) A^T \left[Q_1, Q_{21}^T, Q_{31}^T, \ldots, Q_{\beta-1,1}^T, Q_{\beta 1}^T \right] B \end{aligned} \tag{89}$$

where

$$Q_1 \equiv \sum_{t=1}^{\beta-1} (A^{t-1})^T \Xi A^{t-1} + (A^{\beta-1})^T \Gamma A^{\beta-1}, \qquad Q_{\beta 1} \equiv \Gamma A^{\beta-1}, \tag{90a}$$

$$Q_{\tau 1} \equiv \sum_{t=\tau-1}^{\beta-2} (A^{t-\tau+1})^T \Xi A^{t} + (A^{\beta-\tau})^T \Gamma A^{\beta-1}, \qquad \tau = 2, 3, \dots, \beta - 1. \tag{90b}$$

$$H = R^T \tilde{\tilde{\Xi}} R + \tilde{\Phi} = \begin{pmatrix} H_{11} & H_2^T \\ H_2 & H_{(1)} \end{pmatrix}, \tag{91}$$

where

$$H_{11} \equiv B^T Q_1 B + \Phi, \tag{92a}$$

$$H_2^T \equiv B^T \left[Q_{21}^T, Q_{31}^T, \dots, Q_{\beta-1,1}^T, Q_{\beta 1}^T \right] B, \tag{92b}$$

$$H_{(1)} \equiv \begin{bmatrix} H_{22} & H_{32}^T & \cdots & H_{\beta 2}^T \\ H_{32} & H_{33} & & H_{\beta 3}^T \\ \vdots & \vdots & \ddots & \vdots \\ H_{\beta 2} & H_{\beta 3} & \cdots & H_{\beta\beta} \end{bmatrix}, \tag{92c}$$

$$H_{\tau\tau} \equiv B^T \left\{ \sum_{t=\tau-1}^{\beta-2} (A^{t-\tau+1})^T \Xi A^{t-\tau+1} + (A^{\beta-\tau})^T \Gamma A^{\beta-\tau} \right\} B + \Phi, \qquad \tau = 2, \dots, \beta - 1, \tag{92d}$$

$$H_{\tau j} \equiv B \left\{ \sum_{t=\tau-1}^{\beta-2} (A^{t-\tau+1})^T \Xi A^{t-j+1} + (A^{\beta-\tau})^T \Gamma A^{\beta-j} \right\} B, \qquad \tau > j;\ \tau, j = 2, \dots, \beta - 1, \tag{92e}$$

$$H_{\beta\tau} \equiv B^T \Gamma A^{\beta-\tau} B, \qquad \tau = 2, \dots, \beta - 1, \tag{92f}$$

$$H_{\beta\beta} \equiv B^T \Gamma B + \Phi. \tag{92g}$$

For example, consider the case of $\beta = 2$, where Theil's optimal strategy (85′) can be written as

$$\begin{pmatrix} v^*(1) \\ Ev^*(2) \end{pmatrix} = -H_{(\beta=2)}^{-1} \begin{bmatrix} B^T(\Xi + A^T \Gamma A) A x(0) \\ B^T \Gamma A^2 x(0) \end{bmatrix}, \tag{93}$$

where

$$H_{(\beta=2)} \equiv \begin{pmatrix} B^T(\Xi + A^T\Gamma A)B + \Phi & B^TA^T\Gamma B \\ B^T\Gamma AB & B^T\Gamma B + \Phi \end{pmatrix}. \tag{94}$$

Making the inverse of $H_{(\beta=2)}$ in the form of (32) (Murata (1977, p. 9)), we calculate $v^*(1)$ from (93) as

$$v^*(1)_{\beta=2} = -\left[B^T\Gamma_2 B + \Phi\right]^{-1} B^T\Gamma_2 Ax(0), \tag{95}$$

where

$$\Gamma_2 \equiv \Xi + A^T\Gamma\left[A - B(B^T\Gamma B + \Phi)^{-1}B^T\Gamma A\right]. \tag{96}$$

Equation (95) can be found to be equal to the optimal control rule (11) for $t = 1$ in the case of $\beta = 2$. Thus the first-period certainty equivalence strategy in Theorem 6 coincides with the first-period optimal control rule established in Theorem 1 in this case.

Next, we consider the case of $\beta = 3$, and the corresponding relations (89)–(92). Then the Theil's strategy (85′) reduces to

$$\begin{bmatrix} v^*(1) \\ Ev^*(2) \\ Ev^*(3) \end{bmatrix} = -H^{-1}_{(\beta=3)} \begin{bmatrix} B^T\left(\Xi + A^T(\Xi + A^T\Gamma A)A\right)A \\ B^T(\Xi + A^T\Gamma A)A^2 \\ B^T\Gamma A^3 \end{bmatrix} x(0) \tag{97}$$

where

$$H_{(\beta=3)} \equiv \left[\begin{array}{c:c} B^T\left(\Xi + A^T(\Xi + A^T\Gamma A)A\right)B + \Phi & B^TA^T\left[\Xi + A^T\Gamma A, A^T\Gamma\right]B \\ \hdashline B^T(\Xi + A^T\Gamma A)AB & \\ B^T\Gamma A^2B & H_{(\beta=2)} \end{array}\right]. \tag{98}$$

Again we apply the inverse formula (32) of Murata (1977, p. 9) to $H_{(\beta=3)}$ partitioned above, and calculate $v^*(1)$ from (97) as

$$v^*(1)_{\beta=3} = -\left[B^T\Gamma_3 B + \Phi\right]^{-1} B^T\Gamma_3 Ax(0), \tag{99}$$

where

$$\Gamma_3 \equiv \Xi + A^T\Gamma_2\left[A - B\left(B^T\Gamma_2 B + \Phi\right)^{-1}B^T\Gamma_2 A\right]. \tag{100}$$

$v^*(1)_{\beta=3}$ in (99) is easily found to be the same as the optimal control rule (11) for $t = 1$ in the case of $\beta = 3$. Proceeding this way, we may say that for an arbitrary finite horizon β, the first-period optimal control rule (in Theorem 1) is equal to the first-period certainty equivalence strategy (in

Theorem 6) under the parameter specification of (88) and a, b, Ω. The optimal strategy is given by

$$v^*(1)_{\beta=N} = -\left[B^T\Gamma_N B + \Phi\right]^{-1} B^T\Gamma_N Ax(0) \tag{101}$$

for an N-period horizon planning, where Γ_t can be computed iteratively by

$$\Gamma_t = \Xi + A^T\Gamma_{t=1}\left[A - B\left(B^T\Gamma_{t-1}B + \Phi\right)^{-1} B^T\Gamma_{t-1}A\right], \qquad t = 2, \ldots, N \tag{102}$$

with $\Gamma_1 = \Gamma$. When Γ_t is rewritten as $S(N - t + 1)$, (102) becomes formula (12b) with β replaced by N (in Theorem 1), and strategy (101) is nothing but the control rule (11) for $t = 1$. At the beginning of the second period, the same situation arises except that the planning horizon is now $N - 1$ and the known initial state vector is $x(1)$. Thus the second-period strategy on the certainty equivalence principle is determined as

$$v(2) = v^*(1)_{\beta=N-1} = -\left[B^TS(2)B + \Phi\right]^{-1} B^TS(2)Ax(1), \tag{101'}$$

which is exactly the control rule (11) for $t = 2$. The same is true for the later periods. Thus we conclude that the optimal control rule established in Theorem 1 is successive applications of the first-period certainty equivalence strategy by Theil under the special parameter specification mentioned above. (For the same problem in a slightly different context, see Norman (1974).)

Two final comments are in order. First, the certainty equivalence principle should not be confused with the separation principle in Theorem 3, and a clear distinction between them is made by Bar-Shalom and Tse (1974) for some general discrete-time system. Secondly, when system coefficients are random, as in Section 5.3, the optimal control rule will not enjoy the certainty equivalence property. Holbrook and Howrey (1978) show that, except under certain special assumptions, the rule for the first period will differ from Theil's strategy, calling for a simple policy model in the presence of parameter uncertainty.

5.5. Optimal Control of Macroeconomic Systems

Buchanan and Norton (1971) said that applications of current optimal control methods in macroeconomics were just beginning, and were a long way from practical application in governmental policymaking. Soon afterward three voluminous works by Pindyck (1973), Chow (1975), and Aoki (1976) came out on macroeconomic applications of the optimal control theory. Of these, Pindyck's is an econometric study of linear systems control and is regarded as a straightforward extension of the pioneering research by Buchanan and Norton (1971). Chow (1975) and Aoki (1976) may be deemed the leading advocates of optimal economic control theories. Though they treat some nonlinear systems control problems, they emphasize the linear cases, which are our present concern. On the other hand, the linear feedback control technique has become prevalent among macroeconomic theorists such as Sargent and Wallace (1975, 1976). We shall deduce some additional optimal control rules, rather intuitively, from the control theorems previously established and apply one of them to a macroeconomic policy problem.

For illustrative purposes, we adopt a simple balanced-budget macroeconomic model described by the following three equations:

$$Y_t = C_t + I_t + G_t \tag{24*a}$$

$$I_t = a_0 + a_1(Y_{t-1} - Y_{t-2}) + u_{1t} \tag{24*b}$$

$$C_t = b_0 + b_1(Y_{t-1} - G_{t-1}) + u_{2t} \tag{24*c}$$

where Y_t is national income, C_t is consumption, I_t is investment, G_t is government expenditure, and u_{it} $(i = 1,2)$ is disturbances, all in time period t; a_0, a_1, b_0, and b_1 are constants. Equation (24*a) is a definitional identity, (24*b) represents an investment behavior based on the acceleration principle, and (24*c) shows a consumption function in the one-period lagged disposable income (income minus taxes) with taxes equalized to government expenditure. Since we assume u_{it} $(i = 1,2)$ to be mutually independent stochastic variables with zero means and finite variances, I_t and C_t are stochastic variables. Substituting (24*b) and (24*c) into (24*a) yields

$$Y_t = \alpha_1 Y_{t-1} + \alpha_2 Y_{t-2} + G_t - b_1 G_{t-1} + \alpha_0 + u_t, \tag{25*}$$

where $\alpha_1 \equiv a_1 + b_1$, $\alpha_2 \equiv -a_1$, $\alpha_0 \equiv a_0 + b_0$, and $u_t \equiv u_{1t} + u_{2t}$. Equation (25*) is easily transformed into a state-space form:

$$x(t) = Ax(t-1) + Bv(t) + c + \xi(t), \tag{103}$$

where we set $v(t) \equiv G_t$,

$$x(t) \equiv \begin{bmatrix} G_t \\ Y_{t-1} \\ Y_t \end{bmatrix}, \qquad A \equiv \begin{bmatrix} 0 & 0 & 0 \\ 0 & 0 & 1 \\ -b_1 & \alpha_2 & \alpha_1 \end{bmatrix},$$
$$B \equiv \begin{bmatrix} 1 \\ 0 \\ 1 \end{bmatrix}, \qquad c \equiv \begin{bmatrix} 0 \\ 0 \\ \alpha_0 \end{bmatrix}, \qquad \xi(t) \equiv \begin{bmatrix} 0 \\ 0 \\ u_t \end{bmatrix}. \tag{26*}$$

(For a more general case of this transformation, refer to Application 1 at the end of Section 5.1.)

Noticing that the state vector $x(t)$ as defined in (26*) includes the control variable G_t as a component, we may formulate the associated objective cost function only in terms of state vector $x(t)$:

$$W = \sum_{t=1}^{\beta} (x(t) - a(t))^T \Xi_t (x(t) - a(t)), \qquad (\beta\text{: time horizon}) \quad (104)$$

where $a(t)$ is a vector of targets for the state vector, and Ξ_t is a symmetric positive semidefinite matrix giving the relative penalties for the squared deviations of various variables (including control variables) from their targets. For the present illustrative example, if Ξ_t and $a(t)$ are set at

$$\Xi_t = \begin{bmatrix} \Phi & 0 & 0 \\ 0 & 0 & 0 \\ 0 & 0 & \Xi \end{bmatrix} \delta^{t-1}, \qquad a(t) = \begin{bmatrix} G_t^* \\ Y_{t-1}^* \\ Y_t^* \end{bmatrix}, \quad (27^*)$$

where Φ, Ξ are assumed to be some positive scalars, δ is a time-discount factor ($0 < \delta < 1$), and G_t^*, Y_t^* are the target values of G_t, Y_t, respectively. Then (104) reduces to

$$W = \sum_{t=1}^{\beta} \left(\Xi(Y_t - Y_t^*)^2 + \Phi(G_t - G_t^*)^2\right)\delta^{t-1}. \quad (28^*)$$

Total cost W in (28*) contains the penalty associated with the deviation in control value G_t from its target G_t^*. Thus, in effect, the cost function (104) can be of more general form than it appears.

Chow (1975, ch. 8) employs an objective loss function of form (104) and state-space system (103) with parameters A, B, and c being time-varying. We shall consider the case where coefficient matrices A, B are constant and that exogenous vector c is time-varying $c(t)$ in system (103): i.e.,

$$x(t) = Ax(t-1) + Bv(t) + c(t) + \xi(t), \quad (105)$$

where ξ is a time-independent random vector having zero mean and a finite constant covariance matrix R:

$$\operatorname{cov}(\xi(t), \xi(s)) = R\delta_{st}, \qquad (\delta_{st}: \text{Kronecker delta}). \quad (106)$$

The deterministic problem (with $\xi(t)$ set at zero) corresponding to the stochastic minimization of the expected value of the objective (104) with respect to control variables $v(t)$ subject to the system (105) is close to that of Theorem 16 in Section 2.2. Hence, in view of the certainty equivalence property of the stochastic minimization, we can deduce the following.

Theorem 7 (Chow, 1975). *The optimal feedback control $v(t)$ to the finite time-horizon problem of minimizing the expectation of W in* (104) *subject to the linear time-invariant system* (105) *with known $x(0)$ is given by*

$$v(t) = -K(t)x(t-1) - k_a(t), \qquad t = 1, 2, \ldots, \beta \quad (107)$$

where

$$K(t) \equiv \left[B^T S(t) B \right]^{-1} B^T S(t) A, \tag{108}$$

$$S(\beta) = \Xi_\beta, \tag{109a}$$

$$S(t-1) = \Xi_{t-1} + L(t)S(t)A, \qquad t = 2, 3, \ldots, \beta \tag{109b}$$

$$L(t) \equiv \left[A - BK(t) \right]^T, \qquad t = 2, 3, \ldots, \beta \tag{110}$$

$$k_a(\beta) \equiv \left[B^T S(\beta) B \right]^{-1} B^T S(\beta)(c(\beta) - a(\beta)), \tag{111a}$$

$$k_a(t) \equiv \left[B^T S(t) B \right]^{-1} B^T \{ S(t)c(t) - h(t) \}, \qquad t = 1, 2, \ldots, \beta - 1 \tag{111b}$$

$$h(t-1) = S(t-1)a(t-1) - L(t)\{ S(t)c(t) - h(t) \} \quad \textit{with} \quad h(\beta) = S(\beta)a(\beta). \tag{112}$$

Remark. If we put $\Xi_\beta = \Gamma$ and $\Xi_t = \Xi$ for $t = 1, \ldots, \beta - 1$ in Theorem 7, then the optimal control $v(t)$ of (107) will be identical with that of (34) in Theorem 16 (in Section 2.2) provided Φ is neglected.

When time horizon is extended to infinity, and accordingly Ξ_t is set equal to $\Xi \delta^{t-1}$ $(0 < \delta < 1)$ in Theorem 7 above, its companion theorem can be derived as follows by utilizing the Turnovsky's Lagrange multiplier method expounded in Section 5.3.

Theorem 8. *For the infinite-horizon problem of minimizing the expected value of*

$$W = \sum_{t=1}^{\infty} \delta^{t-1}(x(t) - a(t))^T \Xi (x(t) - a(t)), \tag{104'}$$

where $0 < \delta < 1$, Ξ *is a symmetric positive semidefinite constant matrix, and* $x(t)$, $a(t)$ *are the same as in* (104), *subject to* (105) *with known* $x(0)$, *the optimal control is given by*

$$v(t) = -Kx(t-1) - k(t), \qquad \textit{for} \quad t = 1, 2, \ldots \tag{107'}$$

where

$$K \equiv (B^T S B)^{-1} B^T S A, \tag{108'}$$

$$S = \Xi + LSA, \tag{109'}$$

$$L \equiv [A - BK]^T, \tag{110'}$$

$$k(t) \equiv (B^T S B)^{-1} B^T [I - L]^{-1} \{ Sc(t) - \Xi a(t) \}. \tag{111'}$$

PROOF. Using the same Lagrange multiplier method as for Theorem 5 in Section 5.3, we choose $v(t)$ so as to minimize the expected cost in any time t (δ^{t-1} can be neglected here):

$$E\left\{(x(t)-a(t))^T\Xi(x(t)-a(t))\right\}. \tag{29*}$$

Assuming a feedback control law of the form

$$v(t) = -Kx(t-1)-k(t), \tag{30*}$$

we substitute (30*) into (105), obtaining

$$x(t)=[A-BK]x(t-1)-(Bk(t)-c(t))+\xi(t), \tag{31*}$$

where we note that $E(x(t-1)\xi^T(t))=0$ in view of (106). Defining the asymptotic expectations $Ex(t)\equiv\mu$ and $E(x(t)x^T(t))\equiv X$, we have from (31*)

$$\mu=[A-BK]\mu-(Bk(t)-c(t)), \tag{32*}$$

and

$$\begin{aligned} X=&[A-BK]x[A-BK]^T+(Bk(t)-c(t))(Bk(t)-c(t))^T+R\\ &-2[A-BK]\mu(Bk(t)-c(t))^T. \end{aligned} \tag{33*}$$

The present objective (29*), with $a^T(t)\Xi a(t)$ being deleted, may be reduced to

$$E\big(x^T(t)\Xi x(t)\big)-2a^T(t)\Xi Ex(t)=\operatorname{tr}(\Xi X)-2a^T(t)\Xi\mu, \tag{29**}$$

since

$$\begin{aligned} E\big(x^T(t)\Xi x(t)\big)&=\mu^T\Xi\mu+\operatorname{tr}(\Xi\operatorname{cov}(x(t)))\\ &=\mu^T\Xi\mu+\operatorname{tr}(\Xi X)-\operatorname{tr}(\Xi\mu\mu^T)\\ &=\operatorname{tr}(\Xi X). \end{aligned}$$

Now our problem is to minimize (29**) subject to (32*) and (33*), and hence the corresponding Lagrangian form becomes

$$\begin{aligned} L(t)=&\operatorname{tr}(\Xi X)-2a^T(t)\Xi\mu\\ &+\operatorname{tr}\Big[S\big\{[A-BK]X[A-BK]^T\\ &\qquad+(Bk(t)-c(t))(Bk(t)-c(t))^T+R-X\\ &\qquad-2[A-BK]\mu(Bk(t)-c(t))^T\big\}\Big]\\ &+2p^T(t)\{[A-BK]\mu-(Bk(t)-c(t))-\mu\}, \end{aligned} \tag{34*}$$

where S is the matrix of Lagrange multipliers associated with (33*) and $p(t)$ is the column vector of Lagrange multipliers associated with (32*). We are required to choose X, K, μ, $k(t)$ to minimize $L(t)$ in (34*), assuming S to be symmetric. Thus differentiating (34*) with respect to X, K, μ, $k(t)$, respectively, and setting the results equal to zero, we have

$$\begin{aligned} 0 &= \frac{\partial L(t)}{\partial X} \\ &= \Xi + [A - BK]^T S [A - BK] - S, \end{aligned} \tag{35*a}$$

$$\begin{aligned} 0 &= \frac{1}{2}\frac{\partial L(t)}{\partial K} \\ &= -B^T S[A - BK]X + B^T S(Bk(t) - c(t))\mu^T \\ &\quad - B^T p(t)\mu^T, \end{aligned} \tag{35*b}$$

$$\begin{aligned} 0 &= \frac{1}{2}\frac{\partial L(t)}{\partial \mu} \\ &= -[A - BK]^T S(Bk(t) - c(t)) + [A - BK - I]^T p(t) \\ &\quad - \Xi a(t), \end{aligned} \tag{35*c}$$

$$\begin{aligned} 0 &= \frac{1}{2}\frac{\partial L(t)}{\partial k(t)} \\ &= B^T S(Bk(t) - c(t)) - B^T S[A - BK]\mu - B^T p(t). \end{aligned} \tag{35*d}$$

Subtracting (35*d) $\cdot\, \mu^T$ from (35*b) yields

$$0 = B^T S[A - BK][\mu\mu^T - X]. \tag{36*}$$

On the assumption that $\mu\mu^T - X$ is nonsingular, we get from (36*)

$$B^T S[A - BK] = 0, \tag{36**}$$

from which (108′) follows. Taking (36**) into account, we obtain (109′) from (35*a), and reduce (35*c), (35*d) to

$$0 = [A - BK]^T Sc(t) + [A - BK - I]^T p(t) - \Xi a(t), \tag{35**c}$$

$$0 = B^T S(Bk(t) - c(t)) - B^T p(t). \tag{35**d}$$

Then, it follows from (35**d) and (35**c), respectively, that

$$k(t) = (B^T SB)^{-1} B^T \{Sc(t) + p(t)\}, \tag{37*}$$

and

$$p(t) = [I - L]^{-1}\{LSc(t) - \Xi a(t)\}. \tag{38*}$$

Finally, substituting (38*) into (37*) and considering $I + [I - L]^{-1}L = [I - L]^{-1}$, we reach (111′). □

Application 4. Applying Theorem 8 above to a modified Sargent–Wallace's macroeconomic model with rational expectations, we examine their assertion on optimal money supply. The only difference between our model below and theirs lies in the time structure concerning interest rates. In particular, our investment is realized dependent on the expected real interest rate with a lag of one period, while theirs depends on the contemporaneous one. Thus our modified Sargent–Wallace model, exclusive of a production capacity relation, is described by the following three linear equations:

$$y_t = a_1 k_{t-1} + a_2(p_t - {}_tp^*_{t-1}) + u_{1t}, \tag{39*a}$$

$$y_t = b_1 k_{t-1} - b_2\left[r_{t-1} - ({}_tp^*_{t-1} - p_{t-1})\right] + b_3 z_t + u_{2t}, \tag{39*b}$$

$$m_t = p_t + c_1 y_t - c_2 r_{t-1} + u_{3t}, \tag{39*c}$$

where

a_i, b_i, c_i	coefficients, assumed positive $(i = 1, 2)$
y_t	logarithm of real output,
p_t	logarithm of the price of output,
m_t	logarithm of the stock of nominal money balances,
r_t	nominal rate of interest,
k_t	measure of production capacity,
z_t	vector of exogenous variables,
${}_tp^*_{t-1}$	logarithm of the output price in period t expected as of the end of $t-1$,
u_{it}	stochastic disturbance having zero mean and a finite variance, $(i = 1, 2, 3)$.

Equation (39*a) is the aggregate supply function which is assumed to depend positively on the existing production capacity and the gap in the actual price over the expected one. Equation (39*b) is the aggregate demand function which depends positively on the production capacity and negatively on the real rate of interest expected in the preceding period, besides some exogenous factors. Equation (39*c) represents the money market equilibrium: i.e., the real money balances $(m_t - p_t)$ are equated to the demand for money $(c_1 y_t - c_2 r_{t-1})$. All the equations mentioned above are subject to additive random disturbances.

Substitution of r_{t-1} from (39*c) into (39*b) gives

$$y_t = b_1 k_{t-1} + b_2\left[\frac{1}{c_2}(m_t - p_t - c_1 y_t) + {}_tp^*_{t-1} - p_{t-1}\right] + b_3 z_t + \left(u_{2t} - \frac{b_2}{c_2}u_{3t}\right). \tag{39**b}$$

Putting (39*a) and (39**b) together into a matrix equation, we can obtain the following state-space representation by premultiplying the equation by the inverse of the coefficient matrix of $(y_t, p_t)^T$:

$$\begin{pmatrix} y_t \\ p_t \end{pmatrix} = -\mu b_2 c_2 \begin{pmatrix} 0 & a_2 \\ 0 & 1 \end{pmatrix}\begin{pmatrix} y_{t-1} \\ p_{t-1} \end{pmatrix} + \mu b_2 \begin{pmatrix} a_2 \\ 1 \end{pmatrix} m_t$$
$$+ \mu \begin{pmatrix} a_2 b_2 (c_2 - 1) \\ a_2(c_2 + b_2 c_1) + b_2 c_2 \end{pmatrix} {}_t p^*_{t-1} + \mu \begin{pmatrix} h_{yt} \\ h_{pt} \end{pmatrix} + \mu \begin{pmatrix} \xi_{yt} \\ \xi_{pt} \end{pmatrix}, \qquad (40^*)$$

where

$$\mu \equiv \left[b_1(1 + a_2 c_1) + a_2 c_2 \right]^{-1},$$
$$h_{yt} \equiv (a_1 b_2 + a_2 b_1 c_2) k_{t-1} + a_2 b_3 c_2 z_t,$$
$$h_{pt} \equiv (c_2(b_1 - a_1) - a_1 b_2 c_1) k_{t-1} + b_3 c_2 z_t,$$
$$\xi_{yt} \equiv b_2 u_{1t} + a_2 c_2 u_{2t} - a_2 b_2 u_{3t},$$
$$\xi_{pt} \equiv c_2 u_{2t} - b_2 u_{3t} - (b_2 c_1 + c_2) u_{1t}.$$

Sargent and Wallace (1975) supposed a loss function of the form:

$$W_0 = \sum_{t=1}^{\infty} \delta^{t-1} \left\{ (y_t, p_t) \begin{pmatrix} K_{11} & 0 \\ 0 & K_{22} \end{pmatrix} \begin{pmatrix} y_t \\ p_t \end{pmatrix} + (y_t, p_t) \begin{pmatrix} K_1 \\ K_2 \end{pmatrix} + \left(\frac{K_1}{2} \right)^2 + \left(\frac{K_2}{2} \right)^2 \right\} \qquad (41^*)$$

with $K_{ii} > 0$ $(i = 1, 2)$, and $0 < \delta < 1$. Equation (41*) is very close to

$$W_1 = \sum_{t=1}^{\infty} \delta^{t-1} \left\{ K_{11}(y_t - y^*)^2 + K_{22}(p_t - p^*)^2 \right\}, \qquad (41^{**})$$

where $y^* \equiv -K_1/(2K_{11})$ and $p^* \equiv -K_2/(2K_{22})$. Thus we may consider the problem of minimizing the expected value of W_1 in (41**) with respect to an instrument m_t subject to system (40*). To this problem, we can apply Theorem 8, and the optimal money supply rule is obtained as follows by virtue of (107′):

$$m_t = H \begin{pmatrix} y_{t-1} \\ p_{t-1} \end{pmatrix} + \gamma(t), \qquad (42^*)$$

where

$$H = (0, c_2), \qquad (43^*)$$

$$\gamma(t) = \gamma_0 + \gamma_{1,t} p^*_{t-1} + \gamma_2 k_{t-1} + \gamma_3 z_t \quad (\gamma_i (i = 0, 1, 2, 3)\text{: constants}), \qquad (44^*)$$

since, in view of (110′),

$$L = -\mu b_2 \left(\begin{pmatrix} 0 & 0 \\ a_2 c_2 & c_2 \end{pmatrix} - \begin{pmatrix} 0 \\ c_2 \end{pmatrix} (a_2, 1) \right) = \begin{pmatrix} 0 & 0 \\ 0 & 0 \end{pmatrix} \qquad (45^*)$$

and hence

$$S \equiv \begin{pmatrix} S_{11} & 0 \\ 0 & S_{22} \end{pmatrix} = \begin{pmatrix} K_{11} & 0 \\ 0 & K_{22} \end{pmatrix}, \tag{46*}$$

$$H \equiv \left[(a_2)^2 K_{11} + K_{22}\right]^{-1}(a_2, 1) S \begin{pmatrix} 0 & a_2 c_2 \\ 0 & c_2 \end{pmatrix}, \tag{43**}$$

$$\gamma(t) \equiv \left[(a_2)^2 K_{11} + K_{22}\right]^{-1}\left(\frac{a_2 K_{11}}{b_2}, \frac{K_{22}}{b_2}\right)\begin{bmatrix} \dfrac{y^*}{\mu} - h_{yt} - \alpha_{1,t}p^*_{t-1} \\ \dfrac{p^*}{\mu} - h_{pt} - \alpha_{2,t}p^*_{t-1} \end{bmatrix}, \tag{44**}$$

where $\alpha_1 = a_2 b_2 (c_2 - 1)$ and $\alpha_2 = a_2 c_2 + a_2 b_2 c_1 + b_2 c_2$. Thus the optimal money supply is rewritten

$$m_t = \gamma_0 + c_2 p_{t-1} + \gamma_{1,t} p^*_{t-1} + \gamma_2 k_{t-1} + \gamma_3 z_t, \tag{42**}$$

which is dependent on the expected price ${}_t p^*_{t-1}$ as well as exogenous variables. So the money supply rule (42**) is not deterministic, contrary to the original Sargent–Wallace case. Assume here the public's expectation to be rational by requiring

$${}_t p^*_{t-1} = E_{t-1} p_t, \tag{47*}$$

where $E_{t-1} p_t$ is the mathematical expectation of p_t conditional on all information available at the end of period $t - 1$. With expectation (47*) and (42**) taken into account in (40*), a pseudo-reduced-form equation for p_t is obtained as

$$p_t = \mu b_2 \gamma_0 + \lambda_1 E_{t-1} p_t + \lambda_2 k_{t-1} + \lambda_3 z_t + \mu \xi_{pt} \tag{48*}$$

where λ_1, λ_2, and λ_3 are some constants. Computing $E_{t-1} p_t$ from (48*),

$$E_{t-1} p_t = \mu b_2 \gamma_0 + \lambda_1 E_{t-1} p_t + \lambda_2 k_{t-1} + \lambda_3 z_t + \mu E_{t-1} \xi_{pt}$$

and subtracting the result from (48*), we get

$$p_t - E_{t-1} p_t = \mu(\xi_{pt} - E_{t-1} \xi_{pt}). \tag{49*}$$

Using (47*) and (49*), we write (39*a) as

$$y_t = a_1 k_{t-1} + a_2 \mu(\xi_{pt} - E_{t-1}\xi_{pt}) + u_{1t}. \tag{50*}$$

If we substitute for y_t from (50*) into (39*b), the real interest rate will become a function of k_{t-1} and exogenous processes. Substituting this function into (39*d), determining production capacity, i.e., the Sargent-Wallace's fourth equation (modified as mentioned above):

$$k_t = d_1 k_{t-1} - d_2\left[r_{t-1} - ({}_t p^*_{t-1} - p_{t-1})\right] + d_3 z_t + u_{4t}, \qquad d_2 > 0, \tag{39*d}$$

we get a difference equation in k driven by exogenous processes. This proves, as Sargent and Wallace assert, that k is an exogenous process, which in turn implies, by virtue of (50*), that y is an exogenous process having a distribution independent of the feedback control rule of the money supply.

In a general control model without rational expectations, the distribution of endogenous variables is subject to the effect of a feedback control rule, as shown below.

Application 5. (Cf. Sargent and Wallace (1976).) Consider a single difference equation with an additive disturbance:

$$x_t = ax_{t-1} + bv_t + c + u_t, \tag{51*}$$

where x_t is a controlled variable, v_t is an instrument, u_t is a serially uncorrelated disturbance having zero mean $Eu_t = 0$ and a finite variance $\operatorname{var}(u_t) = \sigma^2$, and a, b, c are constants. Suppose the authority wants to set v_t so as to minimize the variance of x_t over time around some desired level x^*. This is accomplished by choosing the parameters λ_0 and λ_1 in the feedback rule:

$$v_t = \lambda_0 + \lambda_1 x_{t-1}. \tag{52*}$$

Substituting for v_t from (52*) into (51*) yields

$$x_t = (c + b\lambda_0) + (a + b\lambda_1)x_{t-1} + u_t, \tag{53*}$$

which provides the mean of x_t as

$$Ex_t = (c + b\lambda_0) + (a + b\lambda_1)Ex_{t-1}.$$

Alternatively, assuming $Ex_t = Ex_{t-1} \equiv Ex$, we have

$$Ex = (c + b\lambda_0)/(1 - a - b\lambda_1), \tag{54*}$$

which will be equated to x^* in order to minimize the variance of x_t around x^*. On the other hand, in view of the assumptions

$$Eu_t = 0, \quad \operatorname{var}(u_t) = \sigma^2 \quad \text{and} \quad \operatorname{cov}(u_t, x_{t-1}) = 0,$$

we get from (53*)

$$Ex_t^2 = (c + b\lambda_0)^2 + (a + b\lambda_1)^2 Ex_{t-1}^2 + \sigma^2 + 2(c + b\lambda_0)(a + b\lambda_1)Ex_{t-1}. \tag{55*}$$

Then, it follows from (54*) and (55*) that

$$\begin{aligned} \operatorname{var}(x_t) &= Ex_t^2 - (Ex)^2 \\ &= (a + b\lambda_1)^2\left[Ex_{t-1}^2 - (Ex)^2\right] + \sigma^2, \end{aligned} \tag{56*}$$

since

$$\left(1 - (a + b\lambda_1)^2\right)(Ex)^2 = (c + b\lambda_0)^2 + 2(c + b\lambda_0)(a + b\lambda_1)Ex.$$

On the assumption that $\operatorname{var}(x_t) = \operatorname{var}(x_{t-1}) \equiv \operatorname{var}(x)$, we finally obtain from (56*)

$$\operatorname{var}(x) = \frac{\sigma^2}{1 - (a + b\lambda_1)^2} \tag{57*}$$

This shows that the variance of x_t depends on the control parameter λ_1.

Concluding this section, we mention some related literature. For an econometric application of Theorem 7, refer to Abel (1975) who developed a comparison with the case of parameter uncertainty. Chow (1981) provides a nonlinear extension of our linear optimal feedback control. For adaptive control in macroeconomic systems, the reader should consult Upadhyay (1976), Kendrick (1979), and Bar-Shalom and Wall (1980), among others.

References

Abel, A. B., "A comparison of three control algorithms as applied to the monetarist-fiscalist debate," *Annals of Economic and Social Measurement*, vol. 4, pp. 239–252, 1975.

Aoki, M., *Optimization of Stochastic Systems*. New York: Academic Press, 1967.

——, *Optimal Control and System Theory in Dynamic Economic Analysis*. New York: North-Holland, 1976.

Åström, K. J., *Introduction to Stochastic Control Theory*. New York: Academic Press, 1970.

Bar-Shalom, Y., and Tse, E., "Dual effect, certainty equivalence, and separation in stochastic control," *IEEE Trans. Automatic Control*, vol. AC-19, pp. 494–500, 1974.

Bar-Shalom, Y., and Wall, K. D., "Dual adaptive control and uncertainty effects in macroeconomic systems optimization," *Automatica*, vol. 16, pp. 147–156, 1980.

Bertsekas, D. P., *Dynamic Programming and Stochastic Control*. New York: Academic Press, 1976.

Buchanan, L. F., and Norton, F. E., "Optimal control applications in economic systems," in *Advances in Control Systems*, vol. 8, Leondes, C. T., Ed., New York: Academic Press, 1971.

Chow, G. C., *Analysis and Control of Dynamic Economic Systems*. New York: Wiley, 1975.

——, *Econometric Analysis by Control Methods*. New York: Wiley, 1981.

Duchan, A. I., "A clarification and a new proof of the certainty equivalence theorem," *International Economic Review*, vol. 15, pp. 216–224, 1974.

Holbrook, R. S., and Howrey, E. P., "A comparison of the Chow and Theil optimization procedures in the presence of parameter uncertainty," *International Economic Review*, vol. 19, pp. 749–759, 1978.

Kareken, J. H., Muench, T., and Wallace, N., "Optimal open market strategy: The use of information variables," *American Economic Review*, vol. 63, pp. 156–172, 1973.

Kendrick, D., "Adaptive control of macroeconomic models with measurement error," in *Optimal Control for Econometric Models*. Holly, S., *et al.*, Eds., London: Macmillan, 1979, pp. 204–227.

Meier, L., Larson, R. E., and Tether, A. J., "Dynamic programming for stochastic control of discrete systems," *IEEE Trans. Automatic Control*, vol. AC-16, pp. 767–775, 1971.

Murata, Y., *Mathematics for Stability and Optimization of Economic Systems*. New York: Academic Press, 1977.

Norman, A. L., "On the relationship between linear feedback control and first period certainty equivalence," *International Economic Review*, vol. 15, pp. 209–215, 1974.

Pindyck, R. S., *Optimal Planning for Economic Stabilization*. Amsterdam: North-Holland, 1973.

Root, J. C., "Optimum control of non-Gaussian linear stochastic systems with inaccessible state variables," *SIAM Control*, vol. 7, pp. 317–323, 1969.

Sargent, T. J., and Wallace, N., "'Rational' expectations, the optimal monetary instrument, and the optimal money supply rule," *J. Political Economy*, vol. 83, pp. 241–254, 1975.

Sargent, T. J., and Wallace, N., "Rational expectations and the theory of economic policy," *J. Monetary Economics*, vol. 2, pp. 169–183, 1976.

Simon, H. A., "Dynamic programming under uncertainty with a quadratic criterion function," *Econometrica*, vol. 24, pp. 74–81, 1956.

Theil, H., "A note on certainty equivalence in dynamic planning," *Econometrica*, vol. 25, pp. 346–349, 1957.

——, *Economic Forecasts and Policy*, 2nd ed. Amsterdam: North-Holland, 1961.

——, *Optimal Decision Rules for Government and Industry*. Amsterdam: North-Holland, 1964.

Turnovsky, S. J., "Optimal stabilization policies for stochastic linear systems: The case of correlated multiplicative and additive disturbances," *Review of Economic Studies*, vol. 43, pp. 191–194, 1976.

Turnovsky, S. J., "Optimal control of linear systems with stochastic coefficients and additive disturbances," in *Applications of Control Theory to Economic Analysis*, Pitchford, J. D., and Turnovsky, S. J., Eds. Amsterdam: North-Holland, 1977, pp. 293–335.

Upadhyay, T., "Application of adaptive control to economic stabilization policy," *International J. Systems Science*, vol. 7, pp. 641–650, 1976.

Witsenhausen, H. S., "Separation of estimation and control for discrete time systems," *Proc. IEEE*, vol. 59, pp. 1557–1566, 1971.

CHAPTER 6
Stabilization of Economic Systems under the Government Budget Constraint

As an important application of the optimal control methods just established to contemporary economic problems, we take up the stabilization problem of an economy with government budget deficits financed by money and bonds. Section 6.1 discusses the dynamic processes of an economy whose budget deficits are financed by money and one where they are financed by issuing new bonds. The dynamic property of the money-financed budget proves stable, while the bond-financed case is unstable. In Section 6.2, we examine the stability property of an economy with government budget deficits financed by both money and bonds and find its dynamic movement intrinsically unstable. In Section 6.3, the same problem is examined in the corresponding continuous-time economy, and then a Keynesian policy assignment is incorporated into the economy but fails to stabilize it. Thus in Section 6.4 we try to stabilize the economy by applying the optimal control methods established in Chapter 5, using some numerical examples. A similar numerical stabilization analysis is presented in Section 6.5 for an open economy under the government budget constraint and under fixed and flexible exchange rates.

6.1. Dynamic Process of a Government-Budget Constrained Economy

We began this book by introducing in Section 1.1 a Keynesian economy under the government budget constraint, where price level was treated as a variable and random disturbances appeared. Now we neglect price changes and disturbance factors and are thus concerned with the dynamic process of a fixed-price economy in deterministic circumstances under the government budget constraint. Turnovsky (1977, ch. 4) also presents the dynamics of the government budget constraint. Our analysis is similar, emphasis being placed on finding some systematic formula for the dynamic process.

Our model consists of the three Keynesian equations in period t:

$$[\text{IS}] \qquad Y_t = \alpha(1-\tau)(Y_t + B_{t-1}) + I(R_t) + G_t \tag{1a}$$

$$[\text{LM}] \qquad M_t = L(Y_t, R_t) \tag{1b}$$

$$[\text{Budget}] \qquad \Delta M_t + R_t^{-1}\Delta B_t = G_t - \tau Y_t + (1-\tau)B_{t-1}. \tag{1c}$$

The notations, explained below, are mostly those of Section 1.1:

- Y_t — national income in period t,
- B_{t-1} — interest payments on government bonds outstanding at the end of period $t-1$, paid in period t,
- R_t — rate of interest in period t,
- $I(R_t)$ — private investment in period t, a function of R_t,
- G_t — government expenditure in period t,
- M_t — stock of high-powered money in period t,
- $L(Y_t, R_t)$ — demand for money, a function of Y_t and R_t,
- τ — tax rate,
- α — marginal propensity to consume,
- ΔB_t — $B_t - B_{t-1}$, the increment of interest payments on newly issued government bonds in period t,
- ΔM_t — $M_t - M_{t-1}$, additional supply of money (high-powered money) in period t.

A brief description of each equation is in order. Equation (1a) shows the equality of demand for and supply of goods, such that national income is equal to private consumption, $\alpha(1-\tau)(Y_t + B_{t-1})$, plus investment and government expenditure. Equation (1b) means that money supply equals the demand for money, and we adopt the simplest form of demand function in order to minimize the later computational intricacy. Finally, (1c) is the so-called government budget constraint; that is, the additional money supply and the issuance of new bonds, $R_t^{-1}\Delta B_t$, finance the government budget deficits.

The first-order difference forms of (1a) and (1b) are as follows:

$$(1-\alpha(1-\tau))\Delta Y_t + \nu\Delta R_t = \alpha(1-\tau)\Delta B_{t-1} + \Delta G_t \tag{1'a}$$

$$\beta\Delta Y_t - \mu\Delta R_t = \Delta M_t \tag{1'b}$$

where

$$\nu \equiv -dI/dR > 0, \quad \beta \equiv \partial L/\partial Y > 0, \quad \text{and} \quad \mu \equiv -\partial L/\partial R > 0. \tag{2}$$

The system of difference equations (1′) is solved for ΔY_t and ΔR_t:

$$\begin{pmatrix} \Delta Y_t \\ \Delta R_t \end{pmatrix} = \begin{pmatrix} 1-\alpha(1-\tau) & \nu \\ \beta & -\mu \end{pmatrix}^{-1} \begin{pmatrix} \alpha(1-\tau)\Delta B_{t-1} + \Delta G_t \\ \Delta M_t \end{pmatrix}. \tag{3}$$

Equivalently,

$$\Delta Y_t = (\alpha\mu(1-\tau)\Delta B_{t-1} + \mu\Delta G_t + \nu\Delta M_t)/D \tag{4}$$

$$\Delta R_t = (\alpha\beta(1-\tau)\Delta B_{t-1} + \beta\Delta G_t - (1-\alpha(1-\tau))\Delta M_t)/D \tag{5}$$

where

$$D \equiv \sigma\mu + \beta\nu > 0, \qquad \sigma \equiv 1-\alpha(1-\tau) > 0. \tag{6}$$

Let us assume that $\rho \times 100\%$ of the government budget deficit is financed by money supply, i.e.

$$\Delta M_t = \rho(G_t - \tau Y_t + (1-\tau)B_{t-1}), \qquad 0 \leqslant \rho \leqslant 1. \tag{7}$$

Then, taking $\Delta G_t = G_t - G_{t-1}$ and the like into account, we obtain ΔY_t and ΔR_t as follows by substituting (7) into (4) and (5):

$$\begin{aligned} Y_t = \lambda Y_{t-1} + \lambda\frac{\mu+\nu\rho}{D}G_t - \lambda\frac{\mu}{D}G_{t-1} + \lambda(1-\tau)\frac{\alpha\mu+\nu\rho}{D}B_{t-1} \\ - \lambda(1-t)\frac{\alpha\mu}{D}B_{t-2}, \end{aligned} \tag{8}$$

$$\begin{aligned} D(R_t - R_{t-1}) = \tau\sigma\rho Y_t + (\beta-\sigma\rho)G_t - \beta G_{t-1} \\ + (1-\tau)(\alpha\beta-\sigma\rho)B_{t-1} - \alpha\beta(1-\tau)B_{t-2} \end{aligned} \tag{5′}$$

Again, substituting (8) for Y_t into (5′) yields

$$\begin{aligned} R_t = R_{t-1} + \lambda\frac{\tau\sigma\rho}{D}Y_{t-1} + \lambda\frac{\beta-(\sigma-\tau)\rho}{D}G_t - \lambda\frac{\beta+\tau\rho}{D}G_{t-1} \\ + \lambda(1-\tau)\frac{\alpha\beta-(1-\alpha)\rho}{D}B_{t-1} - \lambda(1-\tau)\frac{\alpha(\beta+\tau\rho)}{D}B_{t-2} \end{aligned} \tag{9}$$

where

$$\lambda \equiv D/(D+\nu\tau\rho), \qquad 0 < \lambda \leqslant 1. \tag{10}$$

We shall confine ourselves to multiplier effects on Y, though R will be involved via indirect effects. Applying an iterative substitution on (8), we get

$$\begin{aligned} Y_t = \lambda^t Y_0 + \sum_{i=1}^{t}\lambda^i\left(\frac{\mu+\nu\rho}{D}G_{t-i+1} - \frac{\mu}{D}G_{t-i}\right) \\ + (1-\tau)\sum_{i=1}^{t}\lambda^i\left(\frac{\alpha\mu+\nu\rho}{D}B_{t-i} - \frac{\alpha\mu}{D}B_{t-i-1}\right). \end{aligned} \tag{11}$$

We consider a once-and-for-all increase in the level of G in period 1, i.e.,

$$G_1 = G_0 + \Delta G_1, \qquad G_1 = G_2 = \cdots = G_t, \qquad \Delta G_1 > 0 \tag{12}$$

and assume the initial government budget is balanced, i.e.,

$$\tau Y_0 = G_0 + (1-\tau)B_{-1}, \qquad \Delta M_0 = \Delta B_0 = 0, \tag{13}$$

which imply

$$\tau Y_0 = G_0 + (1-\tau)B_0 \tag{13'}$$

since $B_0 - B_{-1} = \Delta B_0 = 0$. B_t for $t > 0$ is expressed as

$$B_t = \sum_{j=1}^{t} \Delta B_j + B_0 \qquad \text{for} \quad t = 1, 2, \ldots \tag{14}$$

Substitution of (12) and (13) into (11) yields

$$\begin{aligned} Y_t = {} & \lambda^t Y_0 + (1-\lambda^t)\frac{1}{\tau}G_0 + (1-\lambda^t)\frac{1-\tau}{\tau}B_0 + \left((1-\lambda^t)\frac{1}{\tau} + \lambda^t\frac{\mu}{D}\right)\Delta G_1 \\ & + (1-\tau)\lambda^{t-1}\left(\lambda\frac{\nu\rho}{D} - \frac{\alpha\mu}{D}\right)B_0 - (1-\lambda^t)\frac{1-\tau}{\tau}B_0 \\ & + (1-\tau)\lambda\left\{\frac{\alpha\mu+\nu\rho}{D}B_{t-1} + \left(\lambda\frac{\alpha\mu+\nu\rho}{D} - \frac{\alpha\mu}{D}\right)\right. \\ & \qquad \left.\times\left(B_{t-2} + \lambda B_{t-3} + \cdots + \lambda^{t-3}B_1\right)\right\}. \end{aligned} \tag{11'}$$

Then we calculate the coefficients of B_0 in (11′). Since

$$\begin{aligned} & \frac{\alpha\mu+\nu\rho}{D} + \frac{1}{D}(\lambda(\alpha\mu+\nu\rho) - \alpha\mu)(1 + \lambda + \cdots + \lambda^{t-3}) \\ & \quad = \frac{\alpha\mu+\nu\rho}{D} + \frac{D-\alpha\tau\mu}{\tau D}(1-\lambda^{t-2}), \end{aligned}$$

we have

$$\begin{aligned} & \lambda^{t-1}\frac{1}{D}(\lambda\nu\rho - \alpha\mu) - \frac{1-\lambda^t}{\tau} \\ & \qquad + \lambda\left\{\frac{\alpha\mu+\nu\rho}{D} + \frac{1}{D}(\lambda(\alpha\mu+\nu\rho) - \alpha\mu)\frac{1-\lambda^{t-2}}{1-\lambda}\right\} \\ & \quad = \frac{1}{\tau}(\lambda^t - \lambda^{t-1} + \lambda - 1) + \frac{1}{D}\nu\rho(\lambda^t + \lambda) \\ & \quad = \left(\frac{\lambda-1}{\tau} + \frac{\lambda}{D}\nu\rho\right)(\lambda^{t-1} + 1) = 0. \end{aligned}$$

Hence (11′) reduces to the following, with (13′) and (14) taken into account:

$$\begin{aligned} Y_t = {} & Y_0 + \frac{1}{\tau}\left(1 - \lambda^t\frac{D-\tau\mu}{D}\right)\Delta G_1 + (1-\tau)\lambda\frac{\alpha\mu+\nu\rho}{D}\sum_{j=1}^{t-1}\Delta B_j \\ & + (1-\tau)\left(\lambda\frac{\alpha\mu+\nu\rho}{D} - \frac{\alpha\mu}{D}\right) \\ & \times\left(\lambda\sum_{j=1}^{t-2}\Delta B_j + \lambda^2\sum_{j=1}^{t-3}\Delta B_j + \cdots + \lambda^{t-2}\Delta B_1\right). \end{aligned} \tag{15}$$

This is the general formula for Y_t being dependent on the process of new bond issue which finances $(1-\rho)\times 100\%$ of every government budget deficit.

In the special case where $\rho = 1$, i.e., the case of money-financed deficit $\Delta B_t = 0$ for all t, λ equals $\lambda_1 \equiv D/(D+\tau\nu)$ and (15) reduces to

$$(Y_t)_{\rho=1} = Y_0 + \frac{1}{\tau}\left(1-\lambda_1^t \frac{D-\tau\mu}{D}\right)\Delta G_1 \qquad \text{for} \quad t = 1,2,\ldots . \tag{16}$$

Since

$$D - \tau\mu = (1-\tau)(1-\alpha)\mu + \beta\nu > 0,$$

the level of Y_t increases gradually starting from

$$(Y_1)_{\rho=1} = Y_0 + \frac{\nu+\mu}{D+\tau\nu}\Delta G_1, \tag{16a}$$

approaching, as time passes,

$$(Y_\infty)_{\rho=1} = Y_0 + \frac{1}{\tau}\Delta G_1. \tag{16b}$$

In other words, in the money-financed deficit case, the impact multiplier on Y of ΔG_1 is $(\nu+\mu)/(D+\tau\nu)$, and the long-run multiplier is $1/\tau$, as was originally shown by Christ (1968).

Now we turn to the general case (15) where $0 \leqslant \rho < 1$. In this case, Y_t at $t = 1$ is derived as

$$\begin{aligned} Y_1 &= Y_0 + \frac{1}{\tau}\left(1-\lambda\frac{D-\tau\mu}{D}\right)\Delta G_1 \\ &= Y_0 + \frac{\mu+\nu\rho}{D+\tau\nu\rho}\Delta G_1 \end{aligned} \tag{17a}$$

This Y_1 is smaller than $(Y_1)_{\rho=1}$ in (16a), since $\lambda > \lambda_1$ for $\rho < 1$. Other consecutive Y_t are obtained from (15) for $t = 2$, 3, and 4 as follows:

$$Y_2 = Y_0 + \frac{1}{\tau}\left(1-\lambda^2\frac{D-\tau\mu}{D}\right)\Delta G_1 + (1-\tau)\lambda\frac{\alpha\mu+\nu\rho}{D}\Delta B_1 \tag{17b}$$

$$\begin{aligned} Y_3 = Y_0 &+ \frac{1}{\tau}\left(1-\lambda^3\frac{D-\tau\mu}{D}\right)\Delta G_1 + (1-\tau)\left((\lambda+\lambda^2)\frac{\alpha\mu+\nu\rho}{D} - \lambda\frac{\alpha\mu}{D}\right)\Delta B_1 \\ &+ (1-\tau)\lambda\frac{\alpha\mu+\nu\rho}{D}\Delta B_2 \end{aligned}$$

$$\begin{aligned} Y_4 = Y_0 &+ \frac{1}{\tau}\left(1-\lambda^4\frac{D-\tau\mu}{D}\right)\Delta G_1 \qquad\qquad (17c) \\ &+ (1-\tau)\left((\lambda+\lambda^2+\lambda^3)\frac{\alpha\mu+\nu\rho}{D} - (\lambda+\lambda^2)\frac{\alpha\mu}{D}\right)\Delta B_1 \\ &+ (1-\tau)\left((\lambda+\lambda^2)\frac{\alpha\mu+\nu\rho}{D} - \lambda\frac{\alpha\mu}{D}\right)\Delta B_2 \\ &+ (1-\tau)\lambda\frac{\alpha\mu+\nu\rho}{D}\Delta B_3 \end{aligned} \tag{17d}$$

We show below that these consecutive values of Y_t increase for any nonnegative ρ less than one. Letting the first difference of Y_t be $\Delta Y_t \equiv Y_t - Y_{t-1}$, we have from (17)

$$\Delta Y_1 = \frac{\mu + \nu\rho}{D + \tau\nu\rho}\Delta G_1, \tag{18a}$$

$$\Delta Y_2 = (1-\lambda)\lambda\frac{D - \tau\mu}{D}\Delta G_1 + (1-\tau)\lambda\frac{\alpha\mu + \nu\rho}{D}\Delta B_1, \tag{18b}$$

$$\Delta Y_3 = (1-\lambda)\lambda^2\frac{D - \tau\mu}{D}\Delta G_1 + (1-\tau)\lambda\frac{((1-\alpha)\mu + \beta\nu)\nu\rho}{D(D + \tau\nu\rho)}\Delta B_1$$
$$+ (1-\tau)\lambda\frac{\alpha\mu + \nu\rho}{D}\Delta B_2, \tag{18c}$$

$$\Delta Y_4 = (1-\lambda)\lambda^3\frac{D - \tau\mu}{D}\Delta G_1 + (1-\tau)\lambda\frac{((1-\alpha)\mu + \beta\nu)\nu\rho}{D(D + \tau\nu\rho)}$$
$$\times(\lambda\Delta B_1 + \Delta B_2) + (1-\tau)\lambda\frac{\alpha\mu + \nu\rho}{D}\Delta B_3, \tag{18d}$$

and in general for $t = 2, 3, 4, 5, \ldots$

$$\Delta Y_t = (1-\lambda)\lambda^{t-1}\frac{D - \tau\mu}{D}\Delta G_1 + (1-\tau)\lambda\frac{((1-\alpha)\mu + \beta\nu)\nu\rho}{D(D + \tau\nu\rho)}$$
$$\times\left(\lambda^{t-3}\Delta B_1 + \lambda^{t-4}\Delta B_2 + \cdots + \lambda\Delta B_{t-3} + \Delta B_{t-2}\right)$$
$$+ (1-\tau)\lambda\frac{\alpha\mu + \nu\rho}{D}\Delta B_{t-1} \tag{18e}$$

Thus, as long as ΔG_1 and ΔB_t (for all $t = 1, 2, 3, \ldots$) are positive, ΔY_t in (18) are positive for all $t = 1, 2, 3, \ldots$. In particular, if $\rho = 0$, i.e., if the government budget deficits are all financed by bonds, then the above expressions (18) reduce to

$$(\Delta Y_1)_{\rho=0} = \frac{\mu}{D}\Delta G_1 \tag{18$'$a}$$

$$(\Delta Y_t)_{\rho=0} = (1-\tau)\frac{\alpha\mu}{D}\Delta B_{t-1} \qquad \text{for} \quad t = 2, 3, 4, \ldots, \tag{18$'$e}$$

since λ equals one in view of (10). Equation (18′a) shows the impact multiplier of ΔG_1, and (18′e) implies a direct dependence of ΔY_t upon ΔB_{t-1} in the case of $\rho = 0$.

In order for ΔB_t in the above expressions to be traced back to exogenous variables, first we have, in view of (7) and (1c),

$$\Delta B_t = R_t(1-\rho)(G_t - \tau Y_t + (1-\tau)B_{t-1}), \qquad 0 \leqslant \rho < 1 \tag{19}$$

and then R_t must be calculated by (9) and other means. For example,

$$\Delta B_1 = R_1(1-\rho)(G_1 - \tau Y_1 + (1-\tau)B_0),$$

in which (17a), (12), and (13′) are taken into consideration, yields

$$\Delta B_1 = R_1(1-\rho)\lambda \frac{D-\tau\mu}{D}\Delta G_1 > 0. \tag{20a}$$

Since ΔB_1 is positive, we can easily see that Y_2 in (17b) is larger than Y_1 in (17a). Then, by setting t equal to one in (9) and by taking account of (6), (12), and $B_0 = B_{-1}$, we get

$$\begin{aligned} R_1 &= R_0 + \lambda \frac{\sigma\rho}{D}(\tau Y_0 - G_0 - (1-\tau)B_0) + \lambda \frac{\beta - (\sigma-\tau)\rho}{D}\Delta G_1 \\ &= R_0 + \lambda \frac{\beta - (\sigma-\tau)\rho}{D}\Delta G_1. \end{aligned} \tag{21}$$

Finally, substituting (21) into (20a) yields

$$\Delta B_1 = (1-\rho)\left[R_0\lambda \frac{D-\tau\mu}{D}\Delta G_1 + \lambda^2 \frac{(\beta-(\sigma-\tau)\rho)(D-\tau\mu)}{D^2}(\Delta G_1)^2 \right]. \tag{20b}$$

If $\rho = 0$, (21) and (20a) reduce to (21′) and (20′a) below, respectively.

$$R_1 = R_0 + \frac{\beta}{D}\Delta G_1 \tag{21′}$$

$$\Delta B_1 = R_1 \frac{D-\tau\mu}{D}\Delta G_1. \tag{20′a}$$

Similarly, ΔB_2 and ΔB_3 in the case of $\rho = 0$ are obtained as follows.

$$\begin{aligned} \Delta B_2 &= R_2(G_2 - \tau Y_2 + (1-\tau)B_1) \\ &= \left(R_1 + (1-\tau)\frac{\alpha\beta}{D}\Delta B_1\right)\left(\frac{D-\tau\mu}{D}\Delta G_1 + (1-\tau)\frac{D-\alpha\tau\mu}{D}\Delta B_1\right) \\ &= \Delta B_1 + (1-\tau)\frac{\Delta B_1}{D^2}\left[R_1 D(D-\alpha\tau\mu) + \alpha\beta(D-\tau\mu)\Delta G_1 \right. \\ &\qquad \left. + \alpha\beta(1-\tau)(D-\alpha\tau\mu)\Delta B_1\right] \\ &> \Delta B_1 \end{aligned} \tag{20′b}$$

$$\begin{aligned} \Delta B_3 &= R_3(G_3 - \tau Y_3 + (1-\tau)B_2) \\ &= \left(R^2 + (1-\tau)\frac{\alpha\beta}{D}\Delta B_2\right) \\ &\quad \times \left(\frac{D-\tau\mu}{D}\Delta G_1 + (1-\tau)\frac{D-\alpha\tau\mu}{D}(\Delta B_1 + \Delta B_2)\right) \\ &= \Delta B_2 + (1-\tau)\frac{\Delta B_2}{D^2}\left[R^2 D(D-\alpha\tau\mu) + \alpha\beta(D-\tau\mu)\Delta G_1 \right. \\ &\qquad \left. + \alpha\beta(1-\tau)(D-\alpha\tau\mu)(\Delta B_1 + \Delta B_2)\right] \\ &> \Delta B_2. \end{aligned} \tag{20′c}$$

Thus, in the case of $\rho = 0$, $\Delta B_1 < \Delta B_2 < \Delta B_3 < \cdots$ holds and hence $\Delta Y_2 < \Delta Y_3 < \Delta Y_4 < \cdots$ holds in view of (18′e). This means that the aggregate effective demand increases successively over time by ever increasing amounts, and thus the economy is going to be explosive and unstable, in the case where government budget deficits are all financed by newly issued bonds.

6.2. Instability of an Economy with Government Budget Deficits

In Section 6.1, we demonstrated that an economy with the government budget deficits financed fully by issuing new bonds has an explosive and unstable structure, but we did not show the instability of the case where the budget deficits are partly financed by bonds and partly by money. We now prove the plausibility of the latter type of instability.

The model we use here is the linearized form of model (1) in Section 6.1, i.e.,

$$[\text{IS}] \qquad Y_t = \alpha(1-\tau)(Y_t + B_{t-1}) - \nu R_t + G_t \tag{22a}$$

$$[\text{LM}] \qquad M_t = \beta Y_t - \mu R_t \tag{22b}$$

$$[\text{Budget}] \qquad \Delta M_t + R_t^{-1}\Delta B_t = G_t - \tau Y_t + (1-\tau)B_{t-1} \tag{22c}$$

where ν, β, and μ are defined in (2), and the other notations are the familiar ones employed in model (1). When R_t and G_t are given constants denoted $\bar{R}$ and $\bar{G}$, respectively, we let $\bar{Y}$, $\bar{M}$, and $\bar{B}$ be the corresponding equilibrium values of Y_t, M_t, and B_{t-1}, respectively, so that $\Delta M_t = \Delta B_t = 0$ holds in model (22). These equilibrium values are calculated as

$$\begin{bmatrix} \bar{Y} \\ \bar{B} \\ \bar{M} \end{bmatrix} = \begin{bmatrix} \sigma & -\alpha(1-\tau) & 0 \\ -\beta & 0 & 1 \\ \tau & -(1-\tau) & 0 \end{bmatrix}^{-1} \begin{bmatrix} \bar{G} - \nu\bar{R} \\ -\mu\bar{R} \\ \bar{G} \end{bmatrix} \tag{23}$$

where $\sigma \equiv 1 - \alpha(1-\tau) > 0$. In the sequence, however, this equilibrium will prove not to be stable.

We consider a situation where variables Y_t, M_t, and B_t deviate from the equilibrium values and denote the deviations by the corresponding lowercase letters:

$$y_t \equiv Y_t - \bar{Y}, \quad m_t \equiv M_t - \bar{M}, \quad \text{and} \quad b_t \equiv B_t - \bar{B}.$$

Thus, given $R_t = \bar{R}$ and $G_t = \bar{G}$, the deviation form of model (22) becomes

$$\sigma y_t = \alpha(1-\tau)b_{t-1} \tag{22′a}$$

$$m_t = \beta y_t \tag{22′b}$$

$$\Delta m_t + \Delta b_t / \bar{R} = -\tau y_t + (1-\tau)b_{t-1} \tag{22′c}$$

Taking $\Delta m_t = m_t - m_{t-1}$ and $\Delta b_t = b_t - b_{t-1}$ into account, we rewrite model (22′) as

$$\begin{bmatrix} \sigma & 0 & 0 \\ -\beta & 0 & 1 \\ \tau & \dfrac{1}{\bar{R}} & 1 \end{bmatrix} \begin{bmatrix} y_t \\ b_t \\ m_t \end{bmatrix} = \begin{bmatrix} 0 & \alpha(1-\tau) & 0 \\ 0 & 0 & 0 \\ 0 & 1-\tau+\dfrac{1}{\bar{R}} & 1 \end{bmatrix} \begin{bmatrix} y_{t-1} \\ b_{t-1} \\ m_{t-1} \end{bmatrix}. \tag{22''}$$

Premultiplying (22″) by the inverse of the matrix on its left-hand side yields

$$\begin{bmatrix} y_t \\ b_t \\ m_t \end{bmatrix} = \begin{bmatrix} 0 & \dfrac{\alpha(1-\tau)}{\sigma} & 0 \\ 0 & 1+\dfrac{(1-\tau)(1-\alpha(1+\beta))}{\sigma}\bar{R} & \bar{R} \\ 0 & \dfrac{\beta\alpha(1-\tau)}{\sigma} & 0 \end{bmatrix} \begin{bmatrix} y_{t-1} \\ b_{t-1} \\ m_{t-1} \end{bmatrix}. \tag{24}$$

The stability property of the first-order difference equation (24) is dependent on the roots of its associated characteristic equation:

$$\begin{vmatrix} \lambda & -\dfrac{\alpha(1-\tau)}{\sigma} & 0 \\ 0 & \lambda-1-\dfrac{(1-\tau)(1-\alpha(1+\beta))}{\sigma}\bar{R} & -\bar{R} \\ 0 & -\dfrac{\beta\alpha(1-\tau)}{\sigma} & \lambda \end{vmatrix} = 0 \tag{25}$$

which, in turn, is reduced to

$$\lambda^2 - \left(1 + \frac{(1-\tau)(1-\alpha(1+\beta))}{\sigma}\bar{R}\right)\lambda - \frac{\beta\alpha(1-\tau)}{\sigma}\bar{R} = 0. \tag{26}$$

Here we give a useful proposition on stability conditions of the present problem.

Proposition 1. *Consider the equation*

$$\lambda^2 - a\lambda + b = 0. \tag{1*}$$

In order for each root of (1*) *to be less than one in modulus, inequalities* (2*)–(4*) *must hold*:

$$1 - a + b > 0, \tag{2*}$$

$$1 + a + b > 0, \tag{3*}$$

$$b < 1. \tag{4*}$$

PROOF. In the case where roots λ are real ($a^2 \geqslant 4b$), we have

$$1 > |\lambda| = \tfrac{1}{2}\left|a \pm \sqrt{a^2 - 4b}\,\right|.$$

This implies

$$-2 < a \pm \sqrt{a^2 - 4b} < 2,$$

which includes two inequalities:

$$\sqrt{a^2 - 4b} < 2 - a \quad \text{and} \quad -2 - a < -\sqrt{a^2 - 4b}\,.$$

These reduce, respectively, to (2*) and (3*). In the case where roots λ are complex ($a^2 < 4b$), we have

$$1 > |\lambda|^2 = \tfrac{1}{4}\left(a + i\sqrt{4b - a^2}\,\right)\left(a - i\sqrt{4b - a^2}\,\right) = b,$$

which is nothing but (4*). □

Applying this proposition to (26), we know that inequality requirement (2*) is not fulfilled for the equation. Hence the difference equation (24) has proved not to be stable.

6.3. Instability of an Economy with Government Budget Deficits and Keynesian Policy Assignment (Continuous-Time Case)

When a dynamic Keynesian economy destabilizes itself, some policy package is conventionally assigned to offset undesirable deviations in target variables, as initiated by Mundell (1962). The policy assignment problems have usually been worked out in continuous-time systems. We try to fix appropriate policy assignments to our Keynesian economy with government budget deficits in the continuous-time model, as the counterpart of the previous discrete-time model (22) in Section 6.2. Therefore, the continuous-time system presented below is essentially of the same structure as the previous discrete-time system except that we now consider the wealth effect of the bond stock on the demand for money. Thus our model consists of the following three equations (a similar model is found in Scarth(1979)):

$$[\text{IS}] \qquad Y = \alpha(1 - \tau)(Y + B) + I(R) + G \qquad (27a)$$

$$[\text{LM}] \qquad M = L(Y, B, R) \qquad (27b)$$

$$[\text{Budget}] \qquad \frac{dM}{dt} + R^{-1}\frac{dB}{dt} = G + B - \tau(Y + B) \qquad (27c)$$

where notations are the same as in the model in Section 6.1 except that no time subscript appears. Since t here denotes continuous time, differentials dM/dt and dB/dt replace differences ΔM_t and ΔB_t in the budget equation. The wealth effect of the stock of bonds is indicated by the inclusion of B in the function of liquidity preference L with the assumption that

$$L_B \equiv \frac{\partial L}{\partial B} > 0. \tag{28}$$

Assumption (28) is justified by defining our wealth as B/R. Even if wealth were defined as $M + B/R$, assumption (28) would still hold, and the subsequent analysis would be valid with minor amendments.

We begin by showing the existence of a unique equilibrium solution to the system of equations (27). These equations make a dynamic system with Y, B, and M as endogenous variables for given G and R. An equilibrium of this system is a situation where all the endogenous variables assume constant values and hence $dM/dt = dB/dt = 0$ hold. In order to verify that the system has a unique set of equilibrium values Y^*, B^*, and M^* of Y, B, and M, respectively, for given G and R we need only determine that its Jacobian matrix is a P-matrix by invoking the Gale–Nikaido global univalence theorem. (Refer to Murata (1977, p. 27 and p. 258) for the definition of a P-matrix and for the global univalence theorem, respectively.) By denoting

$$\sigma \equiv 1 - \alpha(1-\tau) > 0 \quad \text{and} \quad L_Y \equiv \partial L/\partial Y > 0,$$

we get the Jacobian matrix as follows:

$$\begin{bmatrix} \sigma & -\alpha(1-\tau) & 0 \\ -\tau & 1-\tau & 0 \\ -L_y & -L_B & 1 \end{bmatrix} \tag{29}$$

which is easily seen to be a P-matrix. Hence system (27) can be solved uniquely for Y, B, and M.

Next we prove that system (27) behaves unstably—namely that the endogenous variables tend to deviate further away from the equilibrium values with time. In order to examine the stability property of our differential system in the neighborhood of the stationary equilibrium, we follow the method of Sohmen and Schneeweiss (1969). First, we take the linear approximation of equations in model (27) about the equilibrium for given G and R:

$$\sigma\Delta Y - \alpha(1-\tau)\Delta B = 0 \tag{27'a}$$

$$L_Y\Delta Y + L_B\Delta B = \Delta M \tag{27'b}$$

$$-\tau\Delta Y + (1-\tau)\Delta B = \frac{dM}{dt} + R^{-1}\frac{dB}{dt}, \tag{27'c}$$

where $\Delta Y \equiv Y - Y^*$, $\Delta B \equiv B - B^*$, and $\Delta M \equiv M - M^*$. Second, we obtain the differential forms of (27a) and (27b) for given G and R:

$$\sigma \frac{dY}{dt} - \alpha(1-\tau)\frac{dB}{dt} = 0 \tag{27''a}$$

$$L_Y \frac{dY}{dt} + L_B \frac{dB}{dt} = \frac{dM}{dt}. \tag{27''b}$$

Third, we express these five equations altogether in matrix form:

$$\left[\begin{array}{cc:ccc} -\alpha(1-\tau) & 0 & 0 & 0 & 0 \\ L_B & -1 & 0 & 0 & 0 \\ \hdashline 1-\tau & 0 & 0 & -R^{-1} & -1 \\ 0 & 0 & \sigma & -\alpha(1-\tau) & 0 \\ 0 & 0 & L_Y & L_B & -1 \end{array}\right] \left[\begin{array}{c} \Delta B \\ \Delta M \\ \hdashline dY/dt \\ dB/dt \\ dM/dt \end{array}\right] = \left[\begin{array}{c} -\sigma \\ -L_Y \\ \hdashline \tau \\ 0 \\ 0 \end{array}\right] \Delta Y. \tag{30}$$

By decomposing (30) as separated by dotted lines and solving each decomposed system, we get

$$\Delta B = \frac{\sigma}{\alpha(1-\tau)} \Delta Y \tag{31a}$$

$$\Delta M = \left(L_Y + \frac{\sigma}{\alpha(1-\tau)} L_B\right) \Delta Y \tag{31b}$$

$$\begin{bmatrix} dY/dt \\ dB/dt \\ dM/dt \end{bmatrix} = \begin{bmatrix} \alpha(1-\tau) \\ \sigma \\ \sigma L_B + \alpha(1-\tau)L_Y \end{bmatrix} \frac{(1-\tau)\Delta B - \tau \Delta Y}{D} \tag{32}$$

where

$$D \equiv \sigma(L_B + R^{-1}) + \alpha(1-\tau)L_Y > 0. \tag{33}$$

Substitution of (31a) and (31b) into (32) yields

$$\frac{dY}{dt} = \gamma \Delta Y \tag{34a}$$

$$\frac{dB}{dt} = \gamma \Delta B \tag{34b}$$

$$\frac{dM}{dt} = \gamma \Delta M, \tag{34c}$$

where $\gamma \equiv (1-\alpha)(1-\tau)/D > 0$. Equation (34) implies that if Y, B, and M deviate from their respective stationary equilibrium values Y^*, B^*, and M^*, then the former diverges further from the latter as time passes.

In order to stabilize the economy around the stationary equilibrium, Keynesians manipulate the values of exogenous variables G and/or R, using them as policy instruments (or control variables). One common method of applying instruments to the present stationary stabilization problem is the so-called policy assignment, i.e., the direct connection of the change in each policy instrument with the deviation of a target variable from its intended value. Let Y and M be target variables and G and R be the corresponding policy instruments, and consider the following policy assignments (referring to Swoboda (1972)):

$$\frac{dG}{dt} = \pi_1 \Delta Y, \qquad \pi_1 < 0, \tag{35a}$$

$$\frac{dR}{dt} = \pi_2 \Delta M, \qquad \pi_2 > 0, \tag{35b}$$

where ΔY and ΔM denote deviations from their respective target values, and π's are constants having the fixed signs. These assignments denote usual Keynesian fiscal and monetary policies.

On the other hand, in view of (25)–(27), the comparative static effects of changes in the values of G and R upon Y, B, and M will be

$$\begin{bmatrix} \sigma & 0 & -\alpha(1-\tau) \\ L_Y & -1 & L_B \\ \tau & 0 & \tau - 1 \end{bmatrix} \begin{bmatrix} \Delta Y \\ \Delta M \\ \Delta B \end{bmatrix} = \begin{bmatrix} 1 & I_R \\ 0 & -L_R \\ 1 & 0 \end{bmatrix} \begin{pmatrix} \Delta G \\ \Delta R \end{pmatrix} \tag{36}$$

where $I_R \equiv dI/dR < 0$ and $L_R \equiv \partial L/\partial R < 0$. Equation (36) is solved for ΔY and ΔM as

$$\Delta Y = \Delta G + \frac{I_R}{1-\alpha} \Delta R \tag{37a}$$

$$\Delta M = (L_Y - L_B)\Delta G + M_R \Delta R, \tag{37b}$$

where

$$M_R \equiv L_R + \left(L_Y + \frac{\tau}{1-\tau} L_B\right) \frac{I_R}{1-\alpha} < 0.$$

Combining (37) with the policy assignment (35) yields

$$\begin{bmatrix} \dfrac{dG}{dt} \\ \dfrac{dR}{dt} \end{bmatrix} = \begin{bmatrix} \pi_1 & \dfrac{\pi_1 I_R}{1-\alpha} \\ \pi_2(L_Y - L_B) & \pi_2 M_R \end{bmatrix} \begin{bmatrix} \Delta G \\ \Delta R \end{bmatrix}. \tag{38}$$

Since the two inequalities

$$\pi_1 + \pi_2 M_R < 0, \tag{39a}$$

and

$$\begin{vmatrix} \pi_1 & \dfrac{\pi_1 I_R}{1-\alpha} \\ \pi_2(L_Y - L_B) & \pi_2 M_R \end{vmatrix} = \pi_1 \pi_2 \left\{ L_R + \frac{L_B I_R}{(1-\alpha)(1-\tau)} \right\} > 0 \tag{39b}$$

hold, we know that the differential system (38) is stable, i.e., G and R converge asymptotically to some stationary values. (For the related stability conditions, see for example Murata (1977, p. 93).)

However, as we show below, variables Y, B, and M behave in an unstable manner when policy assignment (35) is incorporated into our model. The differential forms of equations (25) and (26) are, respectively,

$$\sigma \frac{dY}{dt} - \alpha(1-\tau)\frac{dB}{dt} = I_R \frac{dR}{dt} + \frac{dG}{dt},$$

and

$$L_Y \frac{dY}{dt} + L_B \frac{dB}{dt} - \frac{dM}{dt} = -L_R \frac{dR}{dt},$$

into which (35) are incorporated to obtain

$$\sigma \frac{dY}{dt} - \alpha(1-\tau)\frac{dB}{dt} = \pi_2 I_R \Delta M + \pi_1 \Delta Y, \tag{40a}$$

$$L_Y \frac{dY}{dt} + L_B \frac{dB}{dt} - \frac{dM}{dt} = -\pi_2 L_R \Delta M. \tag{40b}$$

We put together these differential equations with the linear approximation (27′) of equation (27) about given G and R, and express them all in matrix form:

$$\begin{bmatrix} 0 & R^{-1} & 1 \\ \sigma & -\alpha(1-\tau) & 0 \\ L_Y & L_B & -1 \end{bmatrix} \begin{bmatrix} dY/dt \\ dB/dt \\ dM/dt \end{bmatrix} = \begin{bmatrix} -\tau\Delta Y + (1-\tau)\Delta B \\ \pi_1 \Delta Y + \pi_2 I_R \Delta M \\ -\pi_2 L_R \Delta M \end{bmatrix}. \tag{41}$$

The solution of (41) given below will determine the behavior of Y, B, and M. Denoting by A the coefficient matrix on the left-hand side of (41), we have

$$\begin{bmatrix} dY/dt \\ dB/dt \\ dM/dt \end{bmatrix} = A^{-1} \begin{bmatrix} -\tau & 1-\tau & 0 \\ \pi_1 & 0 & \pi_2 I_R \\ 0 & 0 & -\pi_2 L_R \end{bmatrix} \begin{bmatrix} \Delta Y \\ \Delta B \\ \Delta M \end{bmatrix}$$

$$= \frac{1}{|A|} \begin{bmatrix} \pi_1 H - \tau(1-\sigma) & (1-\sigma)(1-\tau) & \pi_2(I_R H - (1-\sigma)L_R) \\ -\sigma\tau - \pi_1 L_Y & (1-\tau)\sigma & -\pi_2 K \\ \pi_1 L_Y R^{-1} - \tau J & (1-\tau)J & \pi_2 R^{-1} K \end{bmatrix} \times \begin{bmatrix} \Delta Y \\ \Delta B \\ \Delta M \end{bmatrix} \tag{42}$$

where

$$\begin{aligned} |A| &\equiv (1-\sigma)L_Y + \sigma H > 0 \\ H &\equiv L_B + R^{-1} > 0 \\ J &\equiv (1-\sigma)L_Y + \sigma L_B > 0 \\ K &\equiv L_Y I_R + \sigma L_R < 0. \end{aligned} \tag{43}$$

System (42) is asymptotically stable if and only if the following three conditions are satisfied (for the related stability conditions, see Murata (1977, pp. 93–94)):

$$(1-\tau)(1-\alpha)+\pi_1 H+\pi_2 R^{-1}K<0, \tag{44a}$$

$$\left| A^{-1}\begin{bmatrix} -\tau & 1-\tau & 0 \\ \pi_1 & 0 & \pi_2 I_R \\ 0 & 0 & -\pi_2 L_R \end{bmatrix} \right| = |A|^{-1}(1-\tau)\pi_1\pi_2 L_R<0, \tag{44b}$$

$$\begin{vmatrix} (1-\tau)(1-\alpha)+\pi_1 H & -\pi_2 K & \pi_2((1-\sigma)L_R-I_R H) \\ (1-\tau)J & \pi_1 H+\pi_2 R^{-1}K-\tau(1-\sigma) & (1-\sigma)(1-\tau) \\ \tau J-\pi_1 L_Y R^{-1} & -\sigma\tau-\pi_1 L_Y & \sigma(1-\tau)+\pi_2 R^{-1}K \end{vmatrix} < 0. \tag{44c}$$

Obviously, inequality (44b) does not hold, and hence (42) behaves unstably. For a similar analysis in an open economy, refer to Turnovsky (1979a).)

6.4. Optimal Control of Economic Systems with Bond-Financed Budget Deficits

In this section, we shall apply optimal control methods established in Chapter 5 to stabilizing the economy with government budget deficits financed by money and bonds, which was found to be intrinsically unstable. The model we consider here has a discrete-time structure as employed in Section 6.2, that is, the following Keynesian model with additive stochastic disturbances:

$$Y_t=\alpha(1-\tau)(Y_t+B_{t-1})-\nu R_t+G_t+u_{1t} \tag{45a}$$

$$M_t=\beta Y_t-\mu R_t+u_{2t} \tag{45b}$$

$$\Delta M_t+R_t^{-1}\Delta B_t=G_t-\tau Y_t+(1-\tau)B_{t-1}+u_{3t}\,. \tag{45c}$$

While (45c) is a reiteration of (1c), equations (45a) and (45b) are the linearized forms of the nonlinear equations (1a) and (1b), all with additive disturbances u_{3t}, u_{1t}, and u_{2t}, respectively, which are assumed to be mutually and serially uncorrelated time-independent random variables having zero means and finite constant variances, i.e.,

$$E(u_{it})=0, \qquad \operatorname{var}(u_{it})=\sigma_i^2 \qquad \text{for} \quad i=1,2,3; \tag{46a}$$

$$\operatorname{cov}(u_{it},u_{js})=0 \qquad \text{for all } t,s;\text{ and for} \quad i\neq j. \tag{46b}$$

Parameters ν, β, and μ are those in (2), and the other notations are the same as in system (1). The one-period difference form of (45b) is

$$\Delta M_t = \beta \Delta Y_t - \mu \Delta R_t + u_{2t} - u_{2t-1} \tag{45'b}$$

into which (45c) is substituted for ΔM_t to result in

$$(\tau + \beta)Y_t - \mu R_t = \beta Y_{t-1} - \mu R_{t-1} + G_t + (1-\tau)B_{t-1} - R_t^{-1}\Delta B_t - u_{2t} + u_{2t-1} + u_{3t}. \tag{45''b}$$

We let $(\overline{Y}, \overline{R}, \overline{B}, \overline{G})$ be a set of stationary values of (Y_t, R_t, B_t, G_t) for system (45) in the case that $\Delta M_t = \Delta B_t = 0$ and disturbances are ignored, and denote by y_t, r_t, b_t, g_t the deviations in Y_t, R_t, B_t, G_t from $\overline{Y}, \overline{R}, \overline{B}, \overline{G}$, respectively. Then the deviation forms of equations (45a) and (45″b) are given as follows (cf. footnote 1 in Chapter 1):

$$y_t = \alpha(1-\tau)(y_t + b_{t-1}) - \nu r_t + g_t + u_{1t} \tag{47a}$$

$$(\tau + \beta)y_t - \mu r_t = \beta y_{t-1} - \mu r_{t-1} + g_t + (1-\tau)b_{t-1} - \overline{R}^{-1}(b_t - b_{t-1}) - u_{2t} + u_{2t-1} + u_{3t}. \tag{47b}$$

These equations can be expressed in matrix form:

$$\begin{pmatrix} \sigma & \nu \\ \tau+\beta & -\mu \end{pmatrix}\begin{pmatrix} y_t \\ r_t \end{pmatrix} = \begin{pmatrix} 0 & 0 \\ \beta & -\mu \end{pmatrix}\begin{pmatrix} y_{t-1} \\ r_{t-1} \end{pmatrix} + \begin{pmatrix} 1 & 0 \\ 1 & -\overline{R}^{-1} \end{pmatrix}\begin{pmatrix} g_t \\ b_t \end{pmatrix} + \begin{pmatrix} \alpha(1-\tau) \\ 1-\tau+\overline{R}^{-1} \end{pmatrix} b_{t-1} + \begin{pmatrix} u_{1t} \\ -u_{2t} + u_{2t-1} + u_{3t} \end{pmatrix} \tag{48}$$

where $\sigma \equiv 1 - \alpha(1-\tau)$. Solving (48) for (y_t, r_t), we might obtain a standard state-space form in the two state variables, but for the sake of the subsequent comparison, we first eliminate variable r_t from system (47). By substituting (47a) for r_t into (47b) we get

$$y_t = a y_{t-1} + \delta_0 g_t + \delta_1 g_{t-1} + \omega_0 b_t + \omega_1 b_{t-1} + \omega_2 b_{t-2} + \eta_t \tag{49}$$

where

$$\begin{gathered} a \equiv (\beta + \sigma\mu/\nu)/\rho_0, \qquad \rho_0 \equiv \tau + \beta + \sigma\mu/\nu, \\ \delta_0 \equiv (1 + \mu/\nu)/\rho_0, \qquad \delta_1 \equiv -\mu/(\nu\rho_0), \qquad \omega_0 \equiv 1/(\overline{R}\rho_0), \\ \omega_1 \equiv (1-\tau)(1 + \alpha\mu/\nu)/\rho_0 + 1/(\overline{R}\rho_0), \qquad \omega_2 \equiv -\alpha(1-\tau)\mu/(\nu\rho_0), \\ \eta_t \equiv \left[(u_{1t} - u_{1t-1})\mu/\nu - (u_{2t} - u_{2t-1} - u_{3t})\right]/\rho_0. \end{gathered} \tag{50}$$

Now we choose as our control variables the changes in G_t, B_t from their previous levels G_{t-1}, B_{t-1}:

$$\Delta g_t \equiv g_t - g_{t-1}, \qquad \Delta b_t \equiv b_t - b_{t-1}. \tag{51}$$

Then equations (49) and (51) are altogether represented in matrix form:

$$x(t) = Ax(t-1) + Hv(t) + \eta(t) \tag{52}$$

where

$$A \equiv \begin{bmatrix} a & 1/\rho_0 & \omega_0+\omega_1 & \omega_2 \\ 0 & 1 & 0 & 0 \\ 0 & 0 & 1 & 0 \\ 0 & 0 & 1 & 0 \end{bmatrix}, \qquad H \equiv \begin{bmatrix} \delta_0 & \omega_0 \\ 1 & 0 \\ 0 & 1 \\ 0 & 0 \end{bmatrix},$$
$$x(t) \equiv \begin{bmatrix} y_t \\ g_t \\ b_t \\ b_{t-1} \end{bmatrix}, \qquad v(t) \equiv \begin{pmatrix} \Delta g_t \\ \Delta b_t \end{pmatrix}, \qquad \eta(t) \equiv \begin{bmatrix} \eta_t \\ 0 \\ 0 \\ 0 \end{bmatrix}. \tag{53}$$

Equation (52) is a standard state-space equation in state variable vector $x(t)$. We are interested in minimizing the expected quadratic costs associated with $x(t)$ and $v(t)$ over a finite time period Υ:

$$E\left\{ x^T(\Upsilon)\Gamma x(\Upsilon) + \sum_{t=1}^{\Upsilon} \left(x^T(t-1)\Xi x(t-1) + v^T(t)\Phi v(t) \right) \right\}, \tag{54}$$

where Γ and Θ are positive semidefinite symmetric matrices and Φ is a positive definite matrix. Our problem is to obtain an optimal control rule for $v(t)$ $(t = 1, 2, \ldots, \Upsilon)$ to minimize the cost function (54) subject to (52) and a given initial condition

$$x(0) = x^0. \tag{55}$$

The composite disturbance term $\eta(t)$ has zero mean and a finite variance, and hence the problem is solved by applying Theorem 1 in Section 5.1 above; formally speaking, the optimal control is

$$v(t) = -K(t)x(t-1) \qquad \text{for} \quad t = 1, 2, \ldots, \Upsilon, \tag{56}$$

where

$$K(t) = \left[\Phi + H^T S(t) H\right]^{-1} H^T S(t) A \tag{57}$$

$$S(\Upsilon) = \Gamma, \tag{58a}$$

$$\begin{aligned} S(t-1) &= A^T S(t)\left[A - HK(t)\right] + \Xi \\ &= A^T\left[S(t) - S(t)H\left(\Phi + H^T S(t) H\right)^{-1} H^T S(t)\right]A + \Xi. \end{aligned} \tag{58b}$$

Suppose that we levy some penalties only for deviations y_t and b_t, and hence Γ and Ξ will now take the following diagonal forms:

$$\Gamma = \begin{bmatrix} 1 & 0 & 0 & 0 \\ 0 & 0 & 0 & 0 \\ 0 & 0 & \gamma & 0 \\ 0 & 0 & 0 & 0 \end{bmatrix}, \qquad \Xi = \begin{bmatrix} 1 & 0 & 0 & 0 \\ 0 & 0 & 0 & 0 \\ 0 & 0 & \xi & 0 \\ 0 & 0 & 0 & 0 \end{bmatrix}, \qquad \gamma, \xi > 0. \tag{59}$$

As for instrument costs, we assume

$$\Phi \equiv \begin{pmatrix} \phi_1 & 0 \\ 0 & \phi_2 \end{pmatrix}, \qquad \phi_1, \phi_2 > 0. \tag{60}$$

Then, $S(t)$ will be calculated backward in time by (58) for $t = \Upsilon, \Upsilon - 1, \dots, 2, 1$. For example, $S(\Upsilon - 1)$ becomes as follows:

$$S(\Upsilon - 1) = \frac{1}{\psi} \begin{bmatrix} a^2 + \psi & a/\rho_0 & a\omega_3 & a\omega_2 \\ a/\rho_0 & 1/\rho_0^2 & \omega_3/\rho_0 & \omega_2/\rho_0 \\ a\omega_3 & \omega_3/\rho_0 & s(\Upsilon - 1)_3 & \omega_2\omega_3 \\ a\omega_2 & \omega_2/\rho_0 & \omega_2\omega_3 & \omega_2\omega_2 \end{bmatrix} \tag{61}$$

where

$$\begin{aligned} \psi &\equiv 1 + \frac{\delta_0^2}{\phi_1} + \frac{\omega_0^2}{\gamma + \phi_2} \\ \omega_3 &\equiv \omega_1 + \frac{\omega_0 \phi_2}{\gamma + \phi_2} \\ s(\Upsilon - 1)_3 &\equiv (\omega_0 + \omega_1)\omega_3 + \xi\psi + \frac{\gamma\left[\omega_0\omega_1\phi_1 + (\phi_1 + \delta_0^2)\phi_2\right]}{\phi_1(\gamma + \phi_2)}. \end{aligned} \tag{62}$$

$K(\Upsilon - 1)$ may be calculated by inserting $S(\Upsilon - 1)$ into (57), but with considerable complications. So we calculate $K(t)$'s numerically for a given set of parameter values, which are now assumed to be

$$\tau = 0.20, \qquad \alpha = 0.85, \qquad \beta = 0.12, \qquad \mu = \nu = 0.4, \qquad \bar{R} = 0.05, \tag{63}$$

and

$$\gamma = \xi = 2, \qquad \phi_1 = 0.5, \qquad \phi_2 = 1, \qquad \Upsilon = 5. \tag{64}$$

From (63), we derive the values of composite parameters

$$\begin{aligned} &\sigma = 0.32, \qquad \rho_0 = 0.64, \qquad a = 0.6875, \qquad \delta_0 = 3.125, \\ &\omega_0 = -31.25, \qquad \omega_1 = 33.5625, \qquad \omega_2 = -1.0625, \qquad \delta_1 = -1.5625. \end{aligned} \tag{65}$$

Using these numerical values, we calculate $K(t)$ for $t = \Upsilon, \Upsilon - 1, \ldots, 2, 1$ as shown next.

$$K(5) = \begin{pmatrix} 0.012 & 0.028 & 0.418 & -0.019 \\ -0.021 & -0.047 & -0.030 & 0.032 \end{pmatrix}$$

$$K(4) = \begin{pmatrix} 0.024 & 0.084 & 0.964 & -0.037 \\ -0.020 & -0.041 & 0.025 & 0.030 \end{pmatrix}$$

$$K(3) = \begin{pmatrix} 0.034 & 0.160 & 1.453 & -0.052 \\ -0.019 & -0.034 & 0.074 & 0.029 \end{pmatrix} \tag{66}$$

$$K(2) = \begin{pmatrix} 0.039 & 0.230 & 1.739 & -0.061 \\ -0.018 & -0.027 & 0.103 & 0.028 \end{pmatrix}$$

$$K(1) = \begin{pmatrix} 0.041 & 0.276 & 1.837 & -0.064 \\ -0.018 & -0.022 & 0.113 & 0.027 \end{pmatrix}.$$

The corresponding optimal control values $v(t)$ will be computed by formula (56) if initial condition x^0 is given. We assume that at the beginning $G(0)$ and $B(0)$ are equal to their target values $\overline{G}$ and $\overline{B}$, respectively, while $Y(0)$ is less than $\overline{Y}$ by some amount, say by 10:

$$x^0 = \begin{bmatrix} y_0 \\ g_0 \\ b_0 \\ b_{-1} \end{bmatrix} = \begin{bmatrix} -10 \\ 0 \\ 0 \\ 0 \end{bmatrix}. \tag{67}$$

Starting from x^0 in (67), we compute $v(1) = (\Delta g_1, \Delta b_1)^T$ by rule (56), and calculate g_1 and b_1 by

$$g_t = g_{t-1} + \Delta g_t, \qquad b_t = b_{t-1} + \Delta b_t \tag{51'}$$

with t replaced by one. Then in view of (49), the computed value of y_1 is obtained as

$$y_t = ay_{t-1} + \delta_0 g_t + \delta_1 g_{t-1} + \omega_0 b_t + \omega_1 b_{t-1} + \omega_2 b_{t-2} \tag{49'}$$

with t replaced by one. Thus we get $x(1) = (y_1, g_1, b_1, b_0)^T$ and employ the formula (56) to have the second-period optimal control $v(2)$. Proceeding in this manner, we compute all the optimal control values $v(t)$ and the associated y_t for $t = 1, 2, \ldots, 5$ as in Table 1.

Table 1. Optimal control values $\Delta g_t, \Delta b_t, y_t$ for $\gamma = \xi = 2$

	$t = 1$	2	3	4	5
Δg_t	0.412	0.216	0.107	0.041	0.007
Δb_t	−0.178	0.029	0.037	0.038	0.038
y_t	−0.027	−0.015	−0.009	−0.005	−0.001

Table 2. Optimal control values $\Delta g_t, \Delta b_t, y_t$ for $\gamma = \xi = 0$

	$t = 1$	2	3	4	5
Δg_t	0.047	0.005	0.003	0.002	0.001
Δb_t	−0.215	−0.013	−0.0067	−0.0067	−0.0069
y_t	−0.0066	−0.0003	−0.0001	−0.0001	−0.0002

This result may be compared with other cases in different numerical examples. First we see the features of the optimal values (in Table 2) obtained in the case where we neglect the cost associated with deviations in bonds B_t from its target, setting $\gamma = \xi = 0$, with the other numericals remaining unchanged.

According to Tables 1 and 2, the changes in G_t and B_t from their previous levels become less drastic in the $\gamma = \xi = 0$ case than in the $\gamma = \xi = 2$ case. In either case the final-period values of Δg_5 and y_5 seem to approach zeros as a whole.

Second, we examine the case that penalty imposition on b_t is shifted backward by one period so that Γ and Ξ now become

$$\Gamma = \begin{bmatrix} 1 & 0 & 0 & 0 \\ 0 & 0 & 0 & 0 \\ 0 & 0 & 0 & 0 \\ 0 & 0 & 0 & \gamma^* \end{bmatrix}, \qquad \Xi = \begin{bmatrix} 1 & 0 & 0 & 0 \\ 0 & 0 & 0 & 0 \\ 0 & 0 & 0 & 0 \\ 0 & 0 & 0 & \xi^* \end{bmatrix}, \qquad \gamma^*, \xi^* > 0. \tag{59'}$$

Setting $\gamma^* = \xi^* = 2$ in (59′) and leaving the other numericals unchanged, we have the optimal values in this case (shown in Table 3) which are slightly different from those in Table 1.

Now, as a comparable form, we turn to our original system (48) having r_t as a state variable and rewrite it with the definition of Δb_t as

$$\begin{bmatrix} \sigma & \nu & 0 \\ \tau + \beta & -\mu & 0 \\ 0 & 0 & 1 \end{bmatrix} \begin{bmatrix} y_t \\ r_t \\ b_t \end{bmatrix} = \begin{bmatrix} 0 & 0 & \alpha(1-\tau) \\ \beta & -\mu & 1-\tau \\ 0 & 0 & 1 \end{bmatrix} \begin{bmatrix} y_{t-1} \\ r_{t-1} \\ b_{t-1} \end{bmatrix}$$

$$+ \begin{bmatrix} 1 & 0 \\ 1 & -\bar{R}^{-1} \\ 0 & 1 \end{bmatrix} \begin{pmatrix} g_t \\ \Delta b_t \end{pmatrix} + \begin{bmatrix} u_{1t} \\ -u_t \\ 0 \end{bmatrix} \tag{68}$$

Table 3. Optimal control values $\Delta g_t, \Delta b_t, y_t$ for $\gamma^* = \xi^* = 2$

	$t = 1$	2	3	4	5
Δg_t	0.392	0.197	0.090	0.025	−0.006
Δb_t	−0.180	0.026	0.033	0.033	0.032
y_t	−0.027	−0.015	−0.010	−0.005	0.001

where u_t represents $u_{2t} - u_{2t-1} - u_{3t}$. Premultiplying (68) by the inverse of the coefficient matrix on the left-hand side, we get the following state-space form.

$$\tilde{x}(t) = A\tilde{x}(t-1) + H\tilde{v}(t) + \tilde{\eta}(t), \tag{69}$$

where

$$A \equiv \begin{bmatrix} \beta/\rho_0 & -\mu/\rho_0 & \omega_0 + \omega_1 \\ -\beta/\rho_1 & \mu/\rho_1 & (1-\tau)\left(\dfrac{\alpha}{\rho_2} - \dfrac{1}{\rho_1}\right) \\ 0 & 0 & 1 \end{bmatrix}, \qquad H \equiv \begin{bmatrix} \delta_0 & \omega_0 \\ \dfrac{1}{\rho_2} - \dfrac{1}{\rho_1} & \dfrac{1}{\overline{R}\rho_1} \\ 0 & 1 \end{bmatrix}, \tag{70}$$

$$\tilde{x}(t) \equiv \begin{bmatrix} y_t \\ r_t \\ b_t \end{bmatrix}, \qquad \tilde{v}(t) \equiv \begin{pmatrix} g_t \\ \Delta b_t \end{pmatrix}, \qquad \tilde{\eta}(t) \equiv \begin{bmatrix} \dfrac{\mu}{\nu\rho_0} u_{1t} - \dfrac{1}{\rho_0} u_t \\ \dfrac{1}{\rho_2} u_{1t} + \dfrac{1}{\rho_1} u_t \\ 0 \end{bmatrix} \tag{71}$$

$$\rho_1 \equiv \nu\rho_0/\sigma, \qquad \rho_2 \equiv \nu\rho_0/(\tau + \beta). \tag{72}$$

We are concerned with minimizing the expected quadratic costs associated with $\tilde{x}(t)$ and $\tilde{v}(t)$ over a finite time period Υ:

$$E\left\{\tilde{x}^T(\Upsilon)\Gamma\tilde{x}(\Upsilon) + \sum_{t=1}^{\Upsilon}\left(\tilde{x}^T(t-1)\Xi\tilde{x}(t-1) + \tilde{v}^T(t)\Phi\tilde{v}(t)\right)\right\}, \tag{73}$$

where we assume

$$\Gamma \equiv \begin{bmatrix} \gamma_1 & 0 & 0 \\ 0 & \gamma_2 & 0 \\ 0 & 0 & \gamma_3 \end{bmatrix}, \qquad \Xi \equiv \begin{bmatrix} \xi_1 & 0 & 0 \\ 0 & \xi_2 & 0 \\ 0 & 0 & \xi_3 \end{bmatrix}, \qquad \Phi \equiv \begin{pmatrix} \phi_1 & 0 \\ 0 & \phi_2 \end{pmatrix} \tag{74}$$

with γ_i, ξ_i, ϕ_i being positive scalars for $i = 1, 2$, and γ_3, ξ_3 being nonnegative. Given an initial condition $\tilde{x}(0)$, we can compute the optimal control

$$\tilde{v}(t) = -K(t)\tilde{x}(t-1) \tag{75}$$

where $K(t)$ is that given by formula (57) with A, H in (70) and Γ, Ξ, Φ in (74).

Table 4. Optimal values $g_t, \Delta b_t, y_t$ and r_t for $\gamma_2 = \xi_2 = 60$

	$t = 1$	2	3	4	5
g_t	0.00109	0.03970	0.03965	0.03963	0.03964
Δb_t	−0.06020	−0.00035	−0.00039	−0.00040	−0.00040
y_t	0.00303	−0.00397	−0.00478	−0.00560	−0.00634
r_t	$\frac{3.06}{10000}$	$\frac{0.91}{10000}$	$\frac{0.46}{10000}$	$\frac{0.004}{10000}$	$\frac{-0.496}{10000}$

Table 5. Optimal values $g_t, \Delta b_t, y_t$ and r_t for $\gamma_2 = \xi_2 = 600$

	$t = 1$	2	3	4	5
g_t	0.00099	0.03967	0.03964	0.03964	0.03967
Δb_t	− 0.06020	− 0.00034	− 0.00039	− 0.00039	− 0.00040
y_t	0.00305	− 0.00396	− 0.00478	− 0.00560	− 0.00635
r_t	$\frac{0.306}{10000}$	$\frac{0.091}{10000}$	$\frac{0.046}{10000}$	$\frac{0.0004}{10000}$	$\frac{-0.0497}{10000}$

Let us compute optimal control values $\tilde{v}(t)$ by using the numerical example given in (63) and

$$\begin{gathered} \Upsilon = 5, \qquad \gamma_1 = \xi_1 = 1, \qquad \gamma_2 = \xi_2 = 60, \\ \gamma_3 = \xi_3 = 2, \qquad \phi_1 = 0.5, \qquad \phi_2 = 1, \end{gathered} \tag{76}$$

starting from the initial condition

$$\tilde{x}(0) = \begin{bmatrix} y_0 \\ r_0 \\ b_0 \end{bmatrix} = \begin{bmatrix} -10.0 \\ 0.01 \\ 0.0 \end{bmatrix}. \tag{77}$$

Table 4 shows the computed optimal control values $\tilde{v}(t) = (g_t, \Delta b_t)$ and the associated state variables y_t and r_t.

Note that in the above computation $\tilde{x}(t)$ is iteratively calculated by formula (69) with its disturbance term being deleted. In a similar manner, we compute optimal control values for the same problem in the case of $\gamma_2 = \xi_2 = 600$ with the other numericals remaining unchanged, and the result is shown in Table 5 where we see a significant improvement in r_t in comparison with Table 4.

6.5. Optimal Control of an Open Economy with Bond-Financed Budget Deficits

A numerical analysis, similar to the preceding Section 6.4, is now performed for optimal control of an open economy under the government budget constraint and under fixed and flexible exchange rates. We introduce external transactions into the previous closed system (45). That is

$$Y_t = \alpha(1-\tau)(Y_t + B_{t-1}) - \nu R_t + G_t + \epsilon \mathbf{E}_t - \zeta Y_t + u_{1t} \tag{78a}$$

$$M_t + \pi J_t = \beta Y_t - \mu R_t + u_{2t} \tag{78b}$$

$$\Delta M_t + R_t^{-1} \Delta B_t = G_t - \tau Y_t + (1-\tau) B_{t-1} + u_{3t} \tag{78c}$$

$$\Delta J_t = \epsilon \mathbf{E}_t - \zeta Y_t + \kappa R_t + u_{4t} \tag{78d}$$

where new notations are as follows (the others are those in (45)):

$\mathbf{E}_t$ exchange rate (price of foreign currency in terms of domestic currency),
J_t level of foreign reserves,
$1-\pi$ sterilization rate of foreign reserves ($0 < \pi \leqslant 1$),
ζ marginal propensity to import ($= \partial Z_t / \partial Y_t > 0$),
κ response ratio of foreign capital inflow to domestic interest rate, ($\kappa \geqslant 0$),
u_{4t} random error with $E(u_{4t}) = 0$ and $\operatorname{var}(u_{4t}) = \sigma_4^2 (< 0)$,

$$\epsilon \equiv \frac{dX_t}{d\mathbf{E}_t} - \frac{\partial Z_t(\mathbf{E}_t, Y_t)}{\partial \mathbf{E}_t} \tag{79}$$

where

X_t volume of exports (in terms of domestic currency),
Z_t volume of imports (in terms of domestic currency).

Note that (79) can be arranged as

$$\epsilon \frac{\mathbf{E}_t}{Z_t} = \xi \frac{X_t}{Z_t} + \eta - 1 \qquad (> 0 \text{ by the so-called Marshall-Lerner condition}), \tag{79'}$$

where

$$\xi \equiv \frac{dX_t}{d\mathbf{E}_t} \frac{\mathbf{E}_t}{X_t} > 0 \qquad \text{(exchange-rate elasticity of export)}$$

$$\eta \equiv -\frac{\partial (Z_t / \mathbf{E}_t)}{\partial \mathbf{E}_t} \frac{\mathbf{E}_t^2}{Z_t} > 0 \qquad \text{(exchange-rate elasticity of import)}.$$

Obviously, $\epsilon \mathbf{E}_t - \zeta Y_t$ is the linearized form of trade balance $X_t(\mathbf{E}_t) - Z_t(\mathbf{E}_t, Y_t)$, and κR_t is that of foreign capital inflow. Thus (78d) represents the overall external transactions balanced by a variation in foreign reserves ΔJ_t, and (78a) shows the Keynesian IS relation involving trade balance. The second term πJ_t on the left-hand side of (78b) gives the foreign reserve component of money stock, meaning that $100 \times \pi$ percent of total foreign currency is converted into domestic currency. Hence total stock of high-powered money can be denoted as

$$H_t \equiv M_t + \pi J_t. \tag{80}$$

Multiplying the current external transactions equation (78d) by π and adding to the government budget constraint equation (78c) yield

$$\Delta H_t + R_t^{-1} \Delta B_t = G_t - \tau' Y_t + (1-\tau) B_{t-1} + \pi\kappa R_t + \pi\epsilon \mathbf{E}_t + u_{5t} \tag{81}$$

where $\tau' \equiv \tau + \pi\zeta$ and $u_{5t} \equiv u_{3t} + \pi u_{4t}$.

Our system is now composed of three equations (78a), (78b), and (81), which may be expressed as

$$\sigma' Y_t + \nu R_t = \alpha(1-\tau)B_{t-1} + G_t + \epsilon \mathbf{E}_t + u_{1t} \tag{82a}$$

$$\beta Y_t - \mu R_t = H_t - u_{2t} \tag{82b}$$

$$\tau' Y_t - \pi\kappa R_t = G_t + (1-\tau)B_{t-1} - R_t^{-1}\Delta B_t + \pi\epsilon \mathbf{E}_t - \Delta H_t + u_{5t} \tag{82c}$$

where $\sigma' \equiv 1 - \alpha(1-\tau) + \zeta > 0$. In a fixed exchange rate case, $\mathbf{E}_t$ is given at $\overline{\mathbf{E}}$, and hence trade balance varies only through changes in Y_t, and $\epsilon\mathbf{E}_t$ becomes equal to a constant, $\epsilon\overline{\mathbf{E}} \equiv \mathcal{E}$ say. If $\overline{\mathbf{E}}$ is raised by $100 \times \mathbf{e}_t$ percent, then $\mathcal{E}$ will also increase by $100 \times \mathbf{e}_t$ percent. In this manner, we may manipulate the dynamics of our system through control of the values of $\overline{\mathbf{E}}$. Thus our system under fixed exchange rate regime can be written as

$$\sigma' Y_t + \nu R_t = \alpha(1-\tau)B_{t-1} + G_t + (1+\mathbf{e}_t)\mathcal{E} + u_{1t} \tag{83a}$$

$$\beta Y_t - \mu R_t = H_t - u_{2t} \tag{83b}$$

$$\tau' Y_t - \pi\kappa R_t = G_t + (1-\tau)B_{t-1} - R_t^{-1}\Delta B_t + (1+\mathbf{e}_t)\pi\mathcal{E} - \Delta H_t + u_{5t} \tag{83c}$$

On the other hand, if the exchange rate floats perfectly, then current external balance ΔJ_t will be zero and H_t may be thought to consist of only domestic component M_t of money stock. At the same time π is put to one. Thus, our perfect floating exchange-rate system is shown as

$$\sigma' Y_t + \nu R_t = \alpha(1-\tau)B_{t-1} + G_t + \epsilon \mathbf{E}_t + u_{1t} \tag{84a}$$

$$\beta Y_t - \mu R_t = M_t - u_{2t} \tag{84b}$$

$$\tau' Y_t - \kappa R_t = G_t + (1-\tau)B_{t-1} - R_t^{-1}\Delta B_t + \epsilon \mathbf{E}_t - \Delta M_t + u_{5t}. \tag{84c}$$

In the real world, the exchange rate does not float perfectly in the above sense, but it floats imperfectly so that current international balance will not be zero; in such a case our previous system (82) will be valid.

We begin with the control of our fixed exchange-rate system (83). The first-order difference form of (83b) is

$$\beta \Delta Y_t - \mu \Delta R_t = \Delta H_t - u_{2t} + u_{2t-1}, \tag{83'b}$$

into which (83c) is substituted for ΔH_t, yielding

$$(\tau' + \beta) Y_t - \mu' R_t = \beta Y_{t-1} - \mu R_{t-1} + G_t + (1-\tau)B_{t-1} - R_t^{-1}\Delta B_t + (1+\mathbf{e}_t)\pi\mathcal{E} - u_t^* \tag{83''b}$$

where $\mu' \equiv \mu + \tau\kappa$ and $u_t^* \equiv u_{2t} - u_{2t-1} - u_{5t}$. As we did in Section 6.4, we derive the following matrix form of (83a) and (83″b) in terms of the deviations of variables from their stationary values (cf. (48)):

$$\begin{pmatrix} \sigma' & \nu \\ \tau' + \beta & -\mu' \end{pmatrix}\begin{pmatrix} y_t \\ r_t \end{pmatrix} = \begin{pmatrix} 0 & 0 \\ \beta & -\mu \end{pmatrix}\begin{pmatrix} y_{t-1} \\ r_{t-1} \end{pmatrix} + \begin{pmatrix} 1 & 0 \\ 1 & -\bar{R}^{-1} \end{pmatrix}\begin{pmatrix} g_t \\ b_t \end{pmatrix}$$
$$+ \begin{pmatrix} \alpha(1-\tau) \\ 1-\tau+\bar{R}^{-1} \end{pmatrix} b_{t-1} + \begin{pmatrix} \mathcal{E} \\ \pi\mathcal{E} \end{pmatrix}\mathbf{e}_t + \begin{pmatrix} u_{1t} \\ -u_t^* \end{pmatrix} \tag{85}$$

where we have considered the fact that exchange-rate deviation rate $\mathbf{e}_t$ would be zero in the stationary state. Here we select government expenditure g_t, new issue of bonds Δb_t and exchange-rate deviation rate $\mathbf{e}_t$ as control variables, while state variables are national income y_t, interest rate r_t and the stock of bonds b_t, all in terms of deviations from the initial stationary values. Thus, rewriting (85) as

$$\begin{bmatrix} \sigma' & \nu & 0 \\ \tau'+\beta & -\mu' & 0 \\ 0 & 0 & 1 \end{bmatrix}\begin{bmatrix} y_t \\ r_t \\ b_t \end{bmatrix} = \begin{bmatrix} 0 & 0 & \alpha(1-\tau) \\ \beta & -\mu & 1-\tau \\ 0 & 0 & 1 \end{bmatrix}\begin{bmatrix} y_{t-1} \\ r_{t-1} \\ b_{t-1} \end{bmatrix}$$
$$+ \begin{bmatrix} 1 & \mathcal{E} & 0 \\ 1 & \pi\mathcal{E} & -\bar{R}^{-1} \\ 0 & 0 & 1 \end{bmatrix}\begin{bmatrix} g_t \\ \mathbf{e}_t \\ \Delta b_t \end{bmatrix} + \begin{bmatrix} u_1 \\ -u_t^* \\ 0 \end{bmatrix}$$

and premultiplying by the inverse of the coefficient matrix on the left-hand side, we get a state-space form

$$x(t) = Ax(t-1) + Hv(t) + \eta(t) \tag{86}$$

where

$$x(t) \equiv \begin{bmatrix} y_t \\ r_t \\ b_t \end{bmatrix}, \qquad A \equiv \begin{bmatrix} \beta/\rho_0' & -\mu/\rho_0' & \omega_0' + \omega_1' \\ -\beta/\rho_1' & \mu/\rho_1' & (1-\tau)\left(\dfrac{\alpha}{\rho_2'} - \dfrac{1}{\rho_1'}\right) \\ 0 & 0 & 1 \end{bmatrix}, \tag{87a}$$

$$v(t) \equiv \begin{bmatrix} g_t \\ \mathbf{e}_t \\ \Delta b_t \end{bmatrix}, \qquad H \equiv \begin{bmatrix} \delta_0' & \mathcal{E}(\pi\delta_0' - (1-\pi)\delta_1') & \omega_0' \\ \dfrac{1}{\rho_2'} - \dfrac{1}{\rho_1'} & \mathcal{E}\left(\dfrac{1}{\rho_2'} - \dfrac{\pi}{\rho_1'}\right) & \dfrac{1}{\bar{R}\rho_1'} \\ 0 & 0 & 1 \end{bmatrix}, \tag{87b}$$

$$\eta(t) \equiv \begin{bmatrix} \dfrac{\mu'}{\nu\rho_0'}u_{1t} - \dfrac{1}{\rho_0'}u_t^* \\ \dfrac{1}{\rho_2'}u_{1t} + \dfrac{1}{\rho_1'}u_t^* \\ 0 \end{bmatrix} \tag{87c}$$

and

$$\rho_0' \equiv \tau' + \beta + \sigma'\mu'/\nu, \qquad \rho_1' \equiv \nu\rho_0'/\sigma', \qquad \rho_2' \equiv \nu\rho_0'/(\tau' + \beta),$$
$$\delta_0' \equiv (1 + \mu'/\nu)/\rho_0', \qquad \delta_1' \equiv -\mu'/(\nu\rho_0'), \qquad \omega_0' \equiv -1/(\bar{R}\rho_0'), \tag{88}$$
$$\omega_1' \equiv (1-\tau)(1 + \alpha\mu'/\nu)/\rho'_0 + 1/(\bar{R}\rho_0').$$

We are concerned with minimizing the expected quadratic costs associated with $x(t)$ and $v(t)$ over the finite time period Υ:

$$E\left\{x^T(\Upsilon)\Gamma x(\Upsilon) + \sum_{t=1}^{\Upsilon}\left(x^T(t-1)\Xi x(t-1) + v^T(t)\Phi v(t)\right)\right\}, \tag{89}$$

where we assume for simplicity

$$\Gamma \equiv \begin{bmatrix} \gamma_1 & 0 & 0 \\ 0 & \gamma_2 & 0 \\ 0 & 0 & \gamma_3 \end{bmatrix}, \qquad \Xi \equiv \begin{bmatrix} \xi_1 & 0 & 0 \\ 0 & \xi_2 & 0 \\ 0 & 0 & \xi_3 \end{bmatrix} \qquad \Phi \equiv \begin{bmatrix} \phi_1 & 0 & 0 \\ 0 & \phi_2 & 0 \\ 0 & 0 & \phi_3 \end{bmatrix} \tag{90}$$

with γ_i, ξ_i, ϕ_i being positive scalars for all i. Given an initial condition $x(0)$, we can compute the optimal control $v(t)$ by the rule (56), where $K(t)$ is that of formula (57) with A, H in (87) and Γ, Ξ, Φ in (90).

Using the numerical data given in (63), (76) and the following additional data:

$$\zeta = 0.05, \qquad \pi = 0.8, \qquad \kappa = 0, \qquad \mathcal{E} = 1, \qquad \phi_3 = 1, \tag{91}$$

we obtain the numerical values of composite parameters

$$\tau' = 0.24, \qquad \mu' = 0.4, \qquad \sigma' = 0.37, \qquad \rho_0' = 0.73, \qquad \rho_1' = 0.789 \tag{92}$$
$$\rho_2' = 0.811, \quad \delta_0' = 2.740, \quad \delta_1' = -1.370, \quad \omega_0' = -27.397, \quad \omega_1' = 29.424$$

and those of matrices A and H:

$$A = \begin{bmatrix} 0.164 & -0.548 & 2.027 \\ -0.152 & 0.507 & -0.175 \\ 0 & 0 & 1 \end{bmatrix}, \qquad H = \begin{bmatrix} 2.740 & 2.466 & -27.397 \\ -0.034 & 0.219 & 25.349 \\ 0 & 0 & 1 \end{bmatrix}. \tag{92'}$$

Based upon these values and the following initial condition

$$x(0) = \begin{bmatrix} y_0 \\ r_0 \\ b_0 \end{bmatrix} = \begin{bmatrix} -10.0 \\ 0.01 \\ 0.0 \end{bmatrix} \tag{93}$$

we compute optimal control values $v(t) = (g_t, \mathbf{e}_t, \Delta b_t)$ and the associated optimal state variables y_t and r_t for $t = 1, 2, 3, 4$, and 5, as shown in Table 6. This table is comparable to Table 4, since the latter contains only two control variables, while in the former we have an additional instrument $\mathbf{e}_t$ that works to improve optimal state variable values as a whole.

Table 6. Optimal control $g_t, \mathbf{e}_t, \Delta b_t$ and state y_t, r_t, b_t

	$t = 1$	2	3	4	5
g_t	0.00480	0.03004	0.02917	0.02832	0.02749
$\mathbf{e}_t$	− 0.00476	0.00978	0.01068	0.01158	0.01250
Δb_t	− 0.06010	− 0.00044	− 0.00049	− 0.00051	− 0.00052
y_t	0.00261	− 0.00295	− 0.00366	− 0.00436	− 0.00502
r_t	0.00029	0.00012	0.00008	0.00004	− 0.00001
b_t	− 0.06010	− 0.06055	− 0.06104	−0.06154	− 0.06207

We turn now to the control of the floating exchange-rate system (84). Before converting the system into a control system, we add to the system a lagged effect of exchange rate on trade balance. That is, we replace the term $\epsilon \mathbf{E}_t$ by $\epsilon_0 \mathbf{E}_t + \epsilon_1 \mathbf{E}_{t-1}$, where

$$\epsilon_0 \equiv \frac{\partial X_t}{\partial \mathbf{E}_t} - \frac{\partial Z_t}{\partial \mathbf{E}_t} > 0 \tag{94a}$$

$$\epsilon_1 \equiv \frac{\partial X_t}{\partial \mathbf{E}_{t-1}} - \frac{\partial Z_t}{\partial \mathbf{E}_{t-1}} > 0 \tag{94b}$$

and we note that X_t and Z_t are assumed to be functions of $\mathbf{E}_t$ and $\mathbf{E}_{t-1}$. (Z_t depends on Y_t as well.) Thus we have a new flexible exchange-rate system (95) instead of (84).

$$\sigma' Y_t + \nu R_t = \alpha(1-\tau)B_{t-1} + G_t + \epsilon_0 \mathbf{E}_t + \epsilon_1 \mathbf{E}_{t-1} + u_{1t} \tag{95a}$$

$$\beta Y_t - \mu R_t = M_t - u_{2t} \tag{95b}$$

$$\tau' Y_t - \kappa R_t = G_t + (1-\tau)B_{t-1} - R_t^{-1}\Delta B_t + \epsilon_0 \mathbf{E}_t + \epsilon_1 \mathbf{E}_{t-1} - \Delta M_t + u_{5t}\,. \tag{95c}$$

In this model, the exchange rate is treated as a state variable, and we denote by $\overline{\mathbf{E}}$ a target value of the exchange rate. Let $\overline{Y}$, $\overline{R}$, $\overline{B}$, $\overline{G}$, and $\overline{M}$ be a set of stationary values of Y_t, R_t, B_t, G_t, and M_t, respectively, consistent with the equilibrium that $\mathbf{E}_t = \mathbf{E}_{t-1} = \overline{\mathbf{E}}$ and $\Delta M_t = \Delta B_t = 0$ are realized in system (95) with u_{it} $(i = 1, 2, 5)$ being neglected.

In order to obtain the deviation form of system (95), we proceed in the same way as we did to reach system (85). The first-order difference form of (95b) is

$$\beta \Delta Y_t - \mu \Delta R_t = \Delta M_t - u_t^* \tag{95'b}$$

into which (95c) is substituted for ΔM_t, yielding

$$\begin{aligned}(\tau' + \beta)Y_t - (\mu + \kappa)R_t - \epsilon_0 \mathbf{E}_t \\ = \beta Y_{t-1} - \mu R_{t-1} + \epsilon_1 \mathbf{E}_{t-1} + G_t + (1-\tau)B_{t-1} - R_t^{-1}\Delta B_t - u_t^*\,.\end{aligned} \tag{95''b}$$

Using the notation

$$\mathbf{e}_t \equiv (\mathbf{E}_t - \overline{\mathbf{E}})/\overline{\mathbf{E}}, \quad \mathcal{E}_0 \equiv \epsilon_0 \overline{\mathbf{E}}, \quad \text{and} \quad \mathcal{E}_1 \equiv \epsilon_1 \overline{\mathbf{E}} \tag{96}$$

and denoting by y_t, r_t, b_t, and g_t, respectively, the deviations in variables Y_t, R_t, B_t, and G_t from their stationary values mentioned above, we can express the system of equations (95a), (95″b), and a definition $b_t = b_{t-1} + \Delta b_t$ in terms of these lower-case letters as

$$\begin{aligned}
&\begin{bmatrix} \sigma' & -\mathcal{E}_0 & 0 \\ \tau' + \beta & -\mathcal{E}_0 & 0 \\ 0 & 0 & 1 \end{bmatrix} \begin{bmatrix} y_t \\ \mathbf{e}_t \\ b_t \end{bmatrix} \\
&\quad = \begin{bmatrix} 0 & \mathcal{E}_1 & \alpha(1-\tau) \\ \beta & \mathcal{E}_1 & 1-\tau \\ 0 & 0 & 1 \end{bmatrix} \begin{bmatrix} y_{t-1} \\ \mathbf{e}_{t-1} \\ b_{t-1} \end{bmatrix} \\
&\quad + \begin{bmatrix} 1 & -\nu & 0 \\ 1 & \mu + \kappa & -\overline{R}^{-1} \\ 0 & 0 & 1 \end{bmatrix} \begin{bmatrix} g_t \\ \Delta r_t \\ \Delta b_t \end{bmatrix} \\
&\quad + \begin{bmatrix} -\nu \\ \kappa \\ 0 \end{bmatrix} r_{t-1} + \begin{bmatrix} u_{1t} \\ -u_t^* \\ 0 \end{bmatrix}.
\end{aligned} \tag{97}$$

Selecting g_t, Δr_t, and Δb_t as control variables, and y_t, $\mathbf{e}_t$, and b_t as state variables, we convert the above system (97) into a state-space form (98), by the premultiplication of the inverse of the coefficient matrix on its left-hand side.

$$\tilde{x}(t) = A\tilde{x}(t-1) + B\tilde{v}(t) + c(t) + \eta(t) \tag{98}$$

where

$$\tilde{x}(t) \equiv \begin{bmatrix} y_t \\ \mathbf{e}_t \\ b_t \end{bmatrix}, \qquad A \equiv \begin{bmatrix} \delta_0^* & 0 & \delta_1^* \\ \sigma'\delta_0^* & -\epsilon_1/\epsilon_0 & \sigma'\delta_1^* - \alpha(1-\tau)/\mathcal{E}_0 \\ 0 & 0 & 1 \end{bmatrix} \tag{99a}$$

$$\tilde{v}(t) \equiv \begin{bmatrix} g_t \\ \Delta r_t \\ \Delta b_t \end{bmatrix}, \qquad B \equiv \begin{bmatrix} 0 & \delta_2^* & \omega_0^* \\ -1/\mathcal{E}_0 & \sigma'\delta_2^* + \nu/\mathcal{E}_0 & \sigma'\omega_0^* \\ 0 & 0 & 1 \end{bmatrix} \tag{99b}$$

$$c(t) \equiv r_{t-1} \begin{bmatrix} \delta_2^* + \delta_3^* \\ \sigma'(\delta_2^* + \delta_3^*) + \nu/\mathcal{E}_0 \\ 0 \end{bmatrix},$$

$$\eta(t) \equiv \frac{1}{\rho_0^*} \begin{bmatrix} -u_t^* - u_1 \\ -\sigma' u_t^* - (\tau' + \beta) u_{1t} \\ 0 \end{bmatrix} \tag{99c}$$

and

$$\begin{aligned} &\rho_0^* \equiv (\tau' + \beta - \sigma')\mathcal{E}_0, \qquad \omega_0^* \equiv -1/(\bar{R}\rho_0^*), \qquad \delta_0^* \equiv \beta/\rho_0^* \\ &\delta_1^* \equiv (1-\tau)(1-\alpha)/\rho_0^*, \qquad \delta_2^* \equiv (\nu + \mu + \kappa)/\rho_0^*, \qquad \delta_3^* \equiv -\mu/\rho_0^*. \end{aligned} \tag{100}$$

Note that $c(t)$ in (98) is an exogenous variable vector.

We want to minimize the expected quadratic loss

$$E\left\{\tilde{x}^T(\Upsilon)\Gamma\tilde{x}(\Upsilon) + \sum_{t=1}^{\Upsilon}\left(\tilde{x}^T(t-1)\Xi\tilde{x}(t-1) + \tilde{v}^T(t)\Phi\tilde{v}(t)\right)\right\} \tag{101}$$

over a finite horizon Υ subject to the system (98). Matrices Γ, Ξ, and Φ are assumed to be those in (90). Given an initial condition $\tilde{x}(0)$, the optimal control of this stochastic problem becomes (102), due to certainty equivalence, in view of Theorem 14 in Section 2.2.

$$\tilde{v}(t) = -K(t)\tilde{x}(t-1) - k(t) \qquad \text{for} \quad t = 1, 2, \ldots, \Upsilon, \tag{102}$$

where

$$K(t) \equiv \left[B^T S(t) B + \Phi\right]^{-1} B^T S(t) A \tag{103}$$

$$S(\Upsilon) = \Gamma \tag{104a}$$

$$S(t-1) = A^T S(t)\left[A - BK(t)\right] + \Xi \tag{104b}$$

$$k(\Upsilon) \equiv \left[B^T \Gamma B + \Phi\right]^{-1} B^T \Gamma c(\Upsilon) \tag{105a}$$

$$\begin{aligned} k(\Upsilon-1) \equiv &\left[B^T S(\Upsilon-1) B + \Phi\right]^{-1} B^T \\ &\times \left\{S(\Upsilon-1)c(\Upsilon-1) + \left[A - BK(\Upsilon)\right]^T \Gamma c(\Upsilon)\right\} \end{aligned} \tag{105b}$$

$$\begin{aligned} k(\Upsilon-2) \equiv &\left[B^T S(\Upsilon-2) B + \Phi\right]^{-1} B^T \\ &\times \Big\{S(\Upsilon-2)c(\Upsilon-2) + \left[A - BK(\Upsilon-1)\right]^T \\ &\times \left\{S(\Upsilon-1)c(\Upsilon-1) + \left[A - BK(\Upsilon)\right]^T \Gamma c(\Upsilon)\right\}\Big\}, \end{aligned} \tag{105c}$$

and so forth.

Using the previous numerical data in (63), (76), (91), and

$$\mathcal{E}_0 = 0.6 \quad \text{and} \quad \mathcal{E}_1 = 0.3, \tag{106}$$

we obtain the following numerical values of composite parameters:

$$\begin{aligned} &\rho_0^* = -0.006, \qquad \omega_0^* = 3333.33, \qquad \delta_0^* = -20, \\ &\delta_1^* = -20, \qquad \delta_2^* = -133.33, \qquad \delta_3^* = 66.67 \end{aligned} \tag{107}$$

and those of A, B, and $c(t)$:

$$A = \begin{bmatrix} -20 & 0 & -20 \\ -7.4 & -0.5 & -8.53 \\ 0 & 0 & 1 \end{bmatrix}, \qquad B = \begin{bmatrix} 0 & -133.33 & 3333.33 \\ -1.67 & -48.66 & 1233.33 \\ 0 & 0 & 1 \end{bmatrix},$$

$$c(t) = r_{t-1} \begin{bmatrix} -66.66 \\ -24.0 \\ 0 \end{bmatrix}. \tag{108}$$

Then, starting from the initial conditions:

$$\tilde{x}(0) \equiv \begin{bmatrix} y_0 \\ \mathbf{e}_0 \\ b_0 \end{bmatrix} = \begin{bmatrix} -10 \\ 0.01 \\ 0 \end{bmatrix} \quad \text{and} \quad r_0 = 0, \tag{109}$$

we can compute optimal control values and the associated state variables.

In order to compute $k(t)$, we need to know $c(\tau)$ for $\tau = t, t+1, \ldots, \Upsilon$ beforehand, and in turn $r_{\tau-1}(= r_0 + \Delta r_1 + \cdots + \Delta r_{\tau-1})$. However, since Δr_t is to be determined as a component of optimal control vector $\tilde{v}(t)$ whose computation involves $k(t)$, we have a contradictory loop. To cut off the loop, we first set $k(t)$ equal to zero for all t and compute $\tilde{v}(t)$ by formula (102), say $\tilde{v}_0(t)$. Second, we set Δr_t equal to the corresponding value in $\tilde{v}_0(t)$ and compute $\tilde{v}(t)$ by (102), say $\tilde{v}_1(t)$. Third, we set Δr_t equal to the corresponding value in $\tilde{v}_1(t)$ and compute $\tilde{v}(t)$ by (102); and so forth. Following this procedure, we can converge to a unique optimal control vector $\tilde{v}(t) = (g_t, \Delta r_t, \Delta b_t)$, as shown in Table 7 together with the associated state variable value $\tilde{x}(t) = (y_t, \mathbf{e}_t, b_t)$, only after three or four iterations.

In comparison with

$$S(5) = \begin{bmatrix} 1 & 0 & 0 \\ 0 & 60 & 0 \\ 0 & 0 & 2 \end{bmatrix},$$

it will be of some interest to see solutions $S(t-1)$ to discrete Riccati

Table 7. Optimal control $g_t, \Delta r_t, \Delta b_t$ and state $y_t, \mathbf{e}_t, b_t$

	$t = 1$	2	3	4	5
g_t	0.00859	0.05368	0.05473	0.05381	0.05123
Δr_t	0.02753	0.01009	0.00430	−0.00097	−0.00575
Δb_t	−0.05890	0.00061	0.00055	0.00042	0.00021
y_t	0.00239	−0.00402	−0.00467	−0.00531	−0.00564
$\mathbf{e}_t$	0.00016	0.00040	0.00041	0.00040	0.00026
b_t	−0.05890	−0.05828	−0.05774	−0.05731	−0.05710

equation (104b) for $t = 5$, 4, 3, and 2:

$$\begin{aligned}
S(4) &= \begin{bmatrix} 1.00126 & -0.00002 & 0.01288 \\ 0.00011 & 60.04044 & 0.09584 \\ 0.01346 & 0.09572 & 4.24535 \end{bmatrix} \\
S(3) &= \begin{bmatrix} 1.00263 & 0.00011 & 0.02750 \\ 0.00022 & 60.04046 & 0.10107 \\ 0.02840 & 0.10103 & 6.52121 \end{bmatrix} \\
S(2) &= \begin{bmatrix} 1.00422 & 0.00018 & 0.03804 \\ 0.00034 & 60.04048 & 0.10626 \\ 0.04368 & 0.10645 & 8.80860 \end{bmatrix} \\
S(1) &= \begin{bmatrix} 0.99980 & 0.00022 & 0.05139 \\ 0.00010 & 60.04049 & 0.11148 \\ 0.05212 & 0.11183 & 11.09510 \end{bmatrix}.
\end{aligned} \tag{110}$$

In concluding this section, we refer to some related literature. Turnovsky (1979b) conducts an optimal control analysis of a continuous-time open economy taking no account of the government budget constraint, while Scarth (1975) is concerned with the stability of a discrete-time open economy in consideration of the government budget constraint. For discussions on equilibrium analysis under flexible exchange-rate systems, we have consulted Krueger (1965), Niehans (1975), Dornbusch (1976), Kouri (1976), and Levin (1980), among others.

References

Christ, C. F., "A simple macroeconomic model with a government budget restraint," *J. Political Economy*, vol. 76, pp. 53–67, 1968.

Dornbusch, R., "Exchange rate expectations and monetary policy," *J. International Economics*, vol. 6, pp. 231–244, 1976.

Kouri, P. J. K., "The exchange rate and the balance of payments in the short run and in the long run: A monetary approach," *Scandinavian J. Economics*, vol. 78, pp. 280–304, 1976.

Krueger, A. O., "The impact of alternative government policies under varying exchange systems," *Quarterly J. Economics*, vol. 79, pp. 195–208, 1965.

Levin, J. H., "The dynamics of stabilization policy under flexible exchange rates: A synthesis of the asset and neo-Keynesian approaches," *Oxford Economic Papers*, vol. 32, pp. 411–427, 1980.

Mundell, R. A., "The appropriate use of monetary and fiscal policy for internal and external stability," *IMF Staff Papers*, vol. 9, pp. 70–79, 1962.

Murata, Y., *Mathematics for Stability and Optimization of Economic Systems*, New York: Academic Press, 1977.

Niehans, J., "Some doubts about the efficacy of monetary policy under flexible exchange rates," *J. International Economics*, vol. 5, pp. 275–281, 1975.

Scarth, W. M., "Fiscal policy and the government budget constraint under alternative exchange-rate systems," *Oxford Economic Papers*, vol. 27, pp. 10–20, 1975.

———, "Bond financed fiscal policy and the problem of instrument instability," *J. Macroeconomics*, vol. 1, pp. 107–117, 1979.

Sohmen, E., and Schneeweiss, H., "Fiscal and monetary policies under alternative exchange rate systems: A correction," *Quarterly J. Economics*, vol. 83, pp. 336–340, 1969.

Swoboda, A. K., "On limited information and the assignment problem," in *Stabilization Policies in Interdependent Economies*, Claassen, E., and Salin, P., Eds., Amsterdam: North-Holland, pp. 101–115, 1972.

Turnovsky, S. J., *Macroeconomic Analysis and Stabilization Policy*. London: Cambridge Univ. Press, 1977.

———, "The dynamics of an open economy with endogenous monetary and fiscal policies," *Weltwirtschaftliches Archiv*, vol. 115, pp. 201–223, 1979a.

———, "Optimal monetary and fiscal policies in an open dynamic economy," *Scandinavian J. Economics*, vol. 81, pp. 400–414, 1979b.

APPENDIX

Differentials of Matrix Traces

Here we assemble all the differentials of matrix traces related to operations conducted in Chapters 3 and 5. In the derivation of the following theorems, the elements of matrix A are assumed to be independent of one another.

Theorem 1a. *Let $A = [a_{ij}]$ and $B = [b_{ij}]$ be $m \times n$ and $n \times m$ matrices, respectively, and let $\Omega = [\omega_{ij}]$ be an $n \times n$ symmetric matrix. Then*

$$\frac{\partial \operatorname{tr}(AB)}{\partial A} = B^T, \tag{1}$$

$$\frac{\partial \operatorname{tr}(A\Omega A^T)}{\partial A} = 2A\Omega. \tag{2}$$

PROOF. Since $\operatorname{tr}(AB) = \sum_{i=1}^{m}\sum_{j=1}^{n} a_{ij}b_{ji}$, we have

$$\frac{\partial \operatorname{tr}(AB)}{\partial a_{ij}} = b_{ji} \Rightarrow (1).$$

Since

$$\operatorname{tr}(A\Omega A^T) = \sum_j \sum_i a_{ij}^2\omega_{jj} + 2\sum\sum_{k<j} \sum_i a_{ik}a_{ij}\omega_{kj},$$

we get

$$\frac{\partial \operatorname{tr}(A\Omega A^T)}{\partial a_{ij}} = 2\sum_k a_{ik}\omega_{kj} \Rightarrow (2). \qquad \square$$

Theorem 2a. *Let $A = [a_{ij}]$ be an $m \times n$ variable matrix, and let B, C be $n \times m$, $m \times m$ constant matrices, respectively. Then*

$$\frac{\partial \operatorname{tr}(CAB)}{\partial A} = C^T B^T, \tag{3}$$

$$\frac{\partial \operatorname{tr}(B^T A^T C^T)}{\partial A} = C^T B^T. \tag{4}$$

Proof. Let $\mathbf{d}$ represent a variation. Since

$$\mathbf{d} \operatorname{tr}(CAB) = \operatorname{tr}\left[\mathbf{d}(CAB)\right] = \operatorname{tr}(C \cdot \mathbf{d}A \cdot B) = \operatorname{tr}(BC \cdot \mathbf{d}A),$$

and since the only nonzero column of $BC(\partial A / \partial a_{ij})$ is the jth one which equals the ith column of BC, we have

$$\begin{aligned} \frac{\partial \operatorname{tr}(CAB)}{\partial a_{ij}} &= \operatorname{tr}\left(BC \frac{\partial A}{\partial a_{ij}}\right) \\ &= \text{the } (j,i)\text{th element of } BC. \end{aligned}$$

Hence

$$\frac{\partial \operatorname{tr}(CAB)}{\partial A} = (BC)^T.$$

Equation (4) follows at once from (3) since

$$\operatorname{tr}(B^T A^T C^T) = \operatorname{tr}(CAB). \qquad \square$$

Theorem 3a. *Let $A = [a_{ij}]$ be an $n \times n$ nonsingular matrix, and let $B = [b_{ij}]$ be an $n \times n$ matrix. Then*

$$\frac{\partial \operatorname{tr}(A^{-1}B)}{\partial A} = -(A^{-1}BA^{-1})^T. \tag{5}$$

Proof. Let A_{rs} be the (r,s)th element of A^{-1}. Then

$$\operatorname{tr}(A^{-1}B) = \sum_r \sum_s A_{rs} b_{sr}.$$

Hence

$$\frac{\partial \operatorname{tr}(A^{-1}B)}{\partial a_{ij}} = \sum_r \sum_s \frac{\partial A_{rs}}{\partial a_{ij}} b_{sr} = \operatorname{tr}\left(\frac{\partial A^{-1}}{\partial a_{ij}} B\right). \tag{6}$$

Since $AA^{-1} = I$, we have

$$\frac{\partial A}{\partial a_{ij}} A^{-1} + A \frac{\partial A^{-1}}{\partial a_{ij}} = 0,$$

and thus

$$\frac{\partial A^{-1}}{\partial a_{ij}} = -A^{-1} \frac{\partial A}{\partial a_{ij}} A^{-1}, \tag{7}$$

which is substituted into (6), yielding

$$\frac{\partial \operatorname{tr}(A^{-1}B)}{\partial a_{ij}} = \operatorname{tr}\left(-A^{-1}\frac{\partial A}{\partial a_{ij}}A^{-1}B\right) = \operatorname{tr}\left(-\frac{\partial A}{\partial a_{ij}}A^{-1}BA^{-1}\right)$$

$$= \text{the } (j,i)\text{th element of } (-A^{-1}BA^{-1}),$$

since the only nonzero row of $(\partial A/\partial a_{ij})C$ for an $n \times n$ matrix C, is the ith one which equals the jth row of C. □

Theorem 4a. *Let $A = [a_{ij}]$ be an $m \times n$ variable matrix. Let B and C be $n \times n$ and $m \times m$ constant symmetric matrices, respectively. Then*

$$\frac{\partial \operatorname{tr}(ABA^TC)}{\partial A} = 2CAB, \tag{8}$$

$$\frac{\partial \operatorname{tr}(CABA^T)}{\partial A} = \frac{\partial \operatorname{tr}(A^TCAB)}{\partial A} = \frac{\partial \operatorname{tr}(BA^TCA)}{\partial A} = 2CAB. \tag{9}$$

PROOF. Let $\mathbf{d}$ represent a variation.

$$\mathbf{d}\operatorname{tr}(ABA^TC) = \operatorname{tr}\left[\mathbf{d}(ABA^TC)\right]. \tag{10}$$

Since

$$\mathbf{d}(ABA^TC) = \mathbf{d}A \cdot BA^TC + A \cdot \mathbf{d}(BA^TC),$$

and since

$$\mathbf{d}(BA^TC) = B \cdot \mathbf{d}A^T \cdot C,$$

we have

$$\operatorname{tr}\left[\mathbf{d}(ABA^TC)\right] = \operatorname{tr}(BA^TC \cdot \mathbf{d}A) + \operatorname{tr}(CAB \cdot \mathbf{d}A^T)$$

$$= \operatorname{tr}(BA^TC \cdot \mathbf{d}A) + \operatorname{tr}(\mathbf{d}A \cdot B^TA^TC^T). \tag{11}$$

By the symmetry of B and C and by (10) and (11),

$$\frac{\partial \operatorname{tr}(ABA^TC)}{\partial a_{ij}} = \operatorname{tr}\left(BA^TC\frac{\partial A}{\partial a_{ij}}\right) + \operatorname{tr}\left(\frac{\partial A}{\partial a_{ij}}BA^TC\right)$$

$$= \text{the } (j,i)\text{th element of } (2BA^TC).$$

Hence

$$\frac{\partial \operatorname{tr}(ABA^TC)}{\partial A} = 2(BA^TC)^T = 2CAB.$$

Equation (9) follows immediately from (8) by virtue of the fact that

$$\operatorname{tr}(CABA^T) = \operatorname{tr}(A^TCAB) = \operatorname{tr}(BA^TCA) = \operatorname{tr}(ABA^TC). \quad \square \tag{12}$$

Theorem 5a. *Let $A = [a_{ij}]$ be an $m \times n$ variable matrix. Let B and C be $n \times n$ and $m \times m$, respectively, constant matrices. Then*

$$\frac{\partial \operatorname{tr}(ABA^TC)}{\partial A} = C^TAB^T + CAB, \tag{13}$$

$$\frac{\partial \operatorname{tr}(CABA^T)}{\partial A} = \frac{\partial \operatorname{tr}(A^TCAB)}{\partial A} = \frac{\partial \operatorname{tr}(BA^TCA)}{\partial A} = C^TAB^T + CAB. \tag{14}$$

PROOF. By (11) and (12)

$$\frac{\partial \operatorname{tr}(ABA^TC)}{\partial a_{ij}} = \operatorname{tr}\left(BA^TC\frac{\partial A}{\partial a_{ij}}\right) + \operatorname{tr}\left(B^TA^TC^T\frac{\partial A}{\partial a_{ij}}\right),$$

and hence

$$\begin{aligned}\frac{\partial \operatorname{tr}(ABA^TC)}{\partial A} &= (BA^TC)^T + (B^TA^TC^T)^T \\ &= C^TAB^T + CAB.\end{aligned}$$

Equation (14) follows immediately from (13) by virtue of (12). □

Theorem 6a. *Let $A = [a_{ij}]$ be an $m \times n$ variable matrix. Let B, C, and D be $n \times n$, $m \times n$, and $n \times m$, respectively, constant matrices. Then*

$$\frac{\partial \operatorname{tr}(DABA^TC)}{\partial A} = CDAB + D^TC^TAB^T. \tag{15}$$

PROOF. Let **d** represent a variation. Since

$$\mathbf{d}\operatorname{tr}(DABA^TC) = \operatorname{tr}\left[\mathbf{d}(DABA^TC)\right],$$

and since

$$\mathbf{d}(DABA^TC) = D \cdot \mathbf{d}A \cdot BA^TC + DAB \cdot \mathbf{d}A^T \cdot C,$$

we have

$$\begin{aligned}\mathbf{d}\operatorname{tr}(DABA^TC) &= \operatorname{tr}(\mathbf{d}A \cdot BA^TCD) + \operatorname{tr}(CDAB \cdot \mathbf{d}A^T) \\ &= \operatorname{tr}(\mathbf{d}A \cdot BA^TCD) + \operatorname{tr}(\mathbf{d}A \cdot B^TA^TD^TC^T).\end{aligned}$$

Then

$$\begin{aligned}\frac{\partial \operatorname{tr}(DABA^TC)}{\partial a_{ij}} &= \operatorname{tr}\left(\frac{\partial A}{\partial a_{ij}}BA^TCD\right) + \operatorname{tr}\left(\frac{\partial A}{\partial a_{ij}}B^TA^TD^TC^T\right) \\ &= \text{the } (j,i)\text{th element of } (BA^TCD + B^TA^TD^TC^T).\end{aligned}$$

Hence

$$\frac{\partial \operatorname{tr}(DABA^TC)}{\partial A} = (BA^TCD + B^TA^TD^TC^T)^T = D^TC^TAB^T + CDAB. \quad \square$$

Author Index

Numbers in italics refer to pages on which the complete references are listed.

A

B

C

D

G

H

J

K

L

M

N

O

P

R

S

T

U

V

W

Y

Subject Index

J

K

L

M

O

P

Q

R

S

T

U

V

other Springer-Verlag titles of interest . . .

Price Effects in Input-Output Relations:
A Theoretical and Empirical Study for the Netherlands 1949-1967
P.M.C. de Boer
1982 / 140 pp. / paper
Lecture Notes in Economics and Mathematical Systems: Volume 201
ISBN 0-387-11550-1

SMS–A Program Package for Simulation and Gaming of
Stochastic Market Processes and Learning Behavior
U. Witt and J. Perske
1982 / 276 pp. / paper
Lecture Notes in Economics and Mathematical Systems: Volume 202
ISBN 0-387-11551-X

Compilation of Input-Output Tables
edited by J. Smolka
1982 / 318 pp. / paper
Lecture Notes in Economics and Mathematical Systems: Volume 203
ISBN 0-387-11553-6